The Practice of Statistics

PUTTING THE PIECES TOGETHER

The Practice of Statistics

PUTTING THE PIECES TOGETHER

JOHN D. SPURRIER

University of South Carolina

Duxbury
Thomson Learning™

Pacific Grove ▪ Albany ▪ Belmont ▪ Boston ▪ Cincinnati ▪ Johannesburg ▪ London
Madrid ▪ Melbourne ▪ Mexico City ▪ New York ▪ Scottsdale ▪ Singapore ▪ Tokyo ▪ Toronto

Acquisitions Editor: *Carolyn Crockett*
Editorial Assistant: *Kimberly Raburn*
Production Editor: *Kelsey McGee*
Manuscript Editor: *Susan Pendleton*
Permissions Editor: *Mary Kay Hancharick*

Interior Design and Typesetting: *Anne Draus,
   Scratchgravel Publishing Services*
Cover Design, Cover and Interior Illustration:
   Jennifer Mackres
Cover Printing, Printing and Binding: *Webcom,
   Limited*

For more information, contact:
DUXBURY
511 Forest Lodge Road
Pacific Grove, CA 93950 USA
www.duxbury.com

Printed in Canada

10 9 8 7 6 5 4 3 2 1

Library of Congress Cataloging-in-Publication Data

Spurrier, John D.
    The practice of statistics : putting the pieces together/ John D. Spurrier.

    p. cm.
    Includes bibliographical references.
    ISBN 0 534-36490-X (text)
    1. Statistics.   I Title.

QA276.S69   1999   99-21479
519.5-dc21

■ ■ ■ ■ ■ ABOUT THE AUTHOR

John D. Spurrier grew up in the small town of Butler, Missouri. He was, and still is, very interested in mathematics and sports. He learned at an early age that he could better understand baseball and basketball games by keeping a score sheet. Although he didn't realize it at the time, he was learning the value of descriptive statistics.

After graduating from high school, he attended the Columbia campus of the University of Missouri. He received a bachelor's degree in mathematics and a master's degree and a doctorate in statistics from that school.

Dr. Spurrier joined the faculty of the University of South Carolina in 1974 and currently holds the rank of Professor in the Department of Statistics. While on the faculty, he has served terms as department chair, assistant department chair, graduate director, and director of the department's statistical consulting service. Dr. Spurrier has taught a wide variety of undergraduate and graduate courses. He has published over 70 papers on the applications and theory of statistics and co-authored a book of laboratory experiences in elementary statistics. He has served as an officer of the Section on Physical and Engineering Sciences and the South Carolina Chapter of the American Statistical Association. Dr. Spurrier is a Fellow of the American Statistical Association and a recipient of the University of South Carolina's Michael J. Mungo Teaching Award.

Dr. Spurrier is married to a statistician and has two children.

To Pam, Katie, and Ryan for their love, enthusiasm, and understanding

Contents

4 **Designing an Experiment to Compare Two Horn Activation Buttons**   **23**

Computing Concepts and Procedures: Random number generation, SAS function PROBT, two-dimensional plot

Mathematical Concepts: Derivative, inverse function

Statistical Concepts: Change of variable, density function, dependent sample design, independent sample design, normality assumption, power of a test, comparing two treatments

Materials Required: None

5 **Using Regression to Predict the Weight of Rocks**   **33**

Computing Concepts and Procedures: SAS PROC PLOT, SAS PROC PRINT, SAS PROC REG, two-dimensional plot

Mathematical Concepts: Ellipsoid, rectangular solid

Statistical Concepts: Regression, residual, *r*-square, scatter plot

Materials Required: For each team, a caliper capable of measuring to the nearest 0.01 inch, a scale capable of measuring to the nearest gram, 20 rocks of various sizes but of similar composition, and muffin pans with holes numbered from 1 to 20. The rocks must have dimensions and weights within caliper and scale capacities

6 **Estimating Variance Components in Tack Measurements**   **43**

Computing Concepts and Procedures: SAS PROC VARCOMP

Mathematical Concepts: System of linear equations

Statistical Concepts: Analysis of variance, factorial design, nested design, random effect, replicate, unbiased estimator, variance component

Materials Required: For each team, 4 nominal 1/2-inch carpet tacks, 3 micrometers capable of measuring to 0.001 inch, 2 objects with a premeasured dimension of less than 1 inch, 12-inch length of masking tape

7 **Classifying Plant Leaves** 55

Computing Concepts and Procedures: SAS PROC DISCRIM, two-dimensional plot

Mathematical Concepts: Linear inequality, matrix operations

Statistical Concepts: Bivariate normal distribution, classification, density function, discriminant analysis, scatter plot, unbiased estimator

Materials Required: For each team, a ruler marked in millimeters

8 **Using a Response Surface to Optimize Product Performance** 71

Computing Concepts and Procedures: SAS PROC GLM, SAS PROC REG, three-dimensional plot

Mathematical Concepts: Maximization, quadratic surface, system of linear equations

Statistical Concepts: Experimental design, factor selection, planned experiment, multiple regression, randomization, response surface

Materials Required: For each team, a balsa wood airplane with moveable wings, a ruler, 4 paper clips, and a 50-foot measuring tape

9 **Modeling Breaking Strength with Dichotomous Data** 89

Computing Concepts and Procedures: Programming Newton's method, SAS PROC LOGISTIC

Mathematical Concepts: Maximization, Newton's method, partial derivative, system of nonlinear equations

Statistical Concepts: Bernoulli trial, dichotomous data, goodness of fit, likelihood function, likelihood ratio test, logistic distribution, logistic regression, maximum likelihood estimation, scatter plot

Materials Required: For each team, 99 two-ply facial tissues with a minimum dimension of at least 8 inches, two 7-inch embroidery hoops, three full 12-ounce soft drink cans, a ruler marked in centimeters, and a 1-ounce egg-shaped fishing weight

Preface

Statistics majors and mathematics majors with an emphasis in statistics receive broad training in mathematics, applied and theoretical statistics, computing, communication, and general education. Unfortunately, these students seldom combine the various skills learned throughout the curriculum (with the modest exceptions of using calculus in mathematical statistics and using a statistical computing package in applied statistics). They also seldom gain hands-on experience by participating in experiments or experience in working in teams. These students graduate with a vast array of knowledge but are often very inexperienced in putting the pieces together. They are not ready to make their greatest possible impact in the workplace. We can and must do better!

The Practice of Statistics: Putting the Pieces Together is presented in two parts. In Part I, students put the pieces together by participating in eleven capstone experiences. Capstone experiences are used in several disciplines as the crowning point of the education. These capstone experiences synthesize knowledge acquired throughout the curriculum.

In each capstone experience, students assume the role of a professional statistician and are asked to use their computing, mathematics, and statistics skills to solve a problem and to communicate the results. All capstone experiences involve data and most require data analysis. Some capstone experiences involve familiar statistical tools, while others lead the students to methods they may not have seen before. The capstone experiences can be done in any order.

Part II contains five chapters detailing strategies for sharpening the vital nonstatistical skills of making effective written and oral reports, creating professional-quality visual aids, serving as a statistical consultant, and finding a job. The most elegant experimental design and statistical analysis have little value if they are not what the client needs or if the statistician cannot communicate the results. The chapters on making reports and on finding a job include checklists that students can use to self-assess their efforts.

It is my experience that students using these materials initially are reluctant to make decisions. As they move through the capstone experiences and the nonstatistical skills chapters, they become much more confident. They learn that they can solve important problems and that they can present these results to others.

There are several approaches for using *The Practice of Statistics: Putting the Pieces Together* in the curriculum. The book can be used as the basis for a three semester-hour course. In addition to participating in the eleven capstone experiences and studying the nonstatistical skills chapters, students should make and critique several oral and written reports. The instructor may wish to supplement these activities with consulting sessions (either live or role-play), with a visit to the career center or placement office, and with guest lectures by practicing statisticians.

A different approach is to include one or more of the experiences in several courses throughout the curriculum. For example, Chapters 1, 2, and 10 involve topics from sampling. Chapters 3, 4, 5, 6, and 8 could be used in a statistical methods or experimental design course. Chapters 7 and 11 could be used in mathematical statistics. Chapters 4 and 9 could be used in statistical computing. Chapter 6 could be used in quality control and improvement.

The approach taken at the University of South Carolina is to use five of the capstone experiences and some of the nonstatistical skills chapters in a one semester-hour course that is required for undergraduate statistics majors and is an elective for graduate statistics majors. We also incorporate additional experiences in other courses.

I thank the Division of Undergraduate Education at the National Science Foundation for providing the support that made this work possible through grant number DUE 9455292. I also thank statisticians Obaid Al-Saidy, Don Edwards, John Grego, Jim Lynch, Richard Madsen, Todd Ogden, Joe Padgett, Walt Piegorsch, Jeff Robinson, Tony Rossini, Pam Spurrier, Lori Thombs, and Web West for reading early drafts of these materials and for suggesting numerous improvements. I appreciate the input of reviewers Thomas Barker, Rochester Institute of Technology; G. Rex Bryce, Brigham Young University; Thomas Johnson, North Carolina State University; Robert Stephenson, Iowa State University; and John Wasik, North Carolina State University. I am also grateful to four experts in other fields for valuable advice: technical writing expert Rhonda Grego, public speaking expert Charles Wilbanks, and human resources experts Bob Crutchfield and Valorie Songer provided valuable feedback on the material in Part II.

I would appreciate feedback regarding the use of these materials. Written comments can be sent to me at the Department of Statistics, University of South Carolina, Columbia, SC 29208 or by e-mail to spurrier@stat.sc.edu.

John D. Spurrier

The Practice of Statistics

PUTTING THE PIECES TOGETHER

PART

I

Capstone Experiences

Statistics majors and mathematics majors with an emphasis in statistics are usually trained in mathematics, applied statistics, mathematical statistics, computing, communications, and general education. Many students think of these areas as separate compartments of knowledge. Practicing statisticians, however, make their impact by combining skills from all these areas to solve problems and to clearly communicate the results. That is, they *put the pieces together*.

In Chapters 1–11, you will practice putting the pieces together by participating in capstone experiences. In each capstone experience, you are placed in the role of a practicing statistician and asked to use your computing, mathematical, and statistical skills to solve a problem, and your oral and written communication skills to report the results of your work. These capstone experiences are designed to simulate the types of problems that statisticians might encounter early in their professional careers. Some of the capstone experiences involve familiar statistical tools, whereas others will lead you to methods that you may not have seen before. Some of the capstone experiences involve the steps leading up to data collection, and others involve data collection and analysis.

These capstone experiences may be quite different from your other educational experiences. As in the working world, there are no answers in the back of the book. Strive for good solutions to the challenges posed by these capstone experiences. Realize that there will often be more than one good solution. Don't let the fact that there might be more than one good solution keep you from making a decision.

1

Preparing Data for Analysis

■ ■ ■ ■ ■ INTRODUCTION

Several important tasks must be done to prepare collected data for analysis, including developing coding protocol and data entry formats, entering the data, and checking the entered data. Precision in these tasks is vital to avoid unnecessary errors and delays. Data entry errors affect data analysis. At times, these errors dramatically change the conclusions of a study. Part of the statistician's job is to ensure that the analysis is done on correct data.

Computing Concepts and Procedures

Coding protocol, data editing, data entry, SAS PROC COMPARE

Mathematical Concepts

None

Statistical Concepts

Five-number summary, frequency table, scatter plot, systematic random sample

Materials Required

None

■ ■ ■ ■ ■ THE SETTING

You are a junior statistician employed by On Time Statistics, a statistical consulting company. Your company is hired by the Sayah Medical Clinic to survey a 1-in-20 systematic random sample of patients coming to the clinic over a six-month period. Surveys are mailed to selected patients two days after their visit. Your boss, Karen Carson, has assigned you the task of preparing the data for analysis. A copy of a completed survey is shown in Figure 1.1. You are told to expect 300 completed surveys.

■ ■ ■ ■ ■ THE BACKGROUND

Preparing data for analysis requires several steps. First, the statistician develops a coding protocol. The coding protocol describes the method for assigning survey identification numbers, a unique number assigned to each completed survey. These numbers allow you to identify a survey if questions arise during data editing or analysis. As we expect 300 completed Sayah Medical Clinic surveys, we could write the identification number 001 at the top of the first completed survey, 002 at the top of the second completed survey, and so on.

The coding protocol also assigns numbers or characters to each expected response to each survey item. It is important to assign a code for the case of "no response"(i.e., missing data). For example, the coding protocol might assign the following numbers to the expected responses to item 5 on the Sayah Medical Clinic survey:

1—Excellent	2—Very good	3—Good	4—Fair
5—Poor	6—Don't remember	9—No response	

Certain "unexpected" responses can be anticipated. For example, the coding protocol should explain what to do if someone marks two doctors on item 4 or marks a response between two of the ordered categories on items 5, 6, 8, 9, or 10. The coding protocol also explains what to do if there is a response not covered by the protocol. Generally, surveys containing responses not covered by the protocol should be returned to you for a coding decision.

The second step in data preparation is to code each completed survey. This begins by writing the identification number at the top of the survey. Then, the correct code numbers or characters are written in the left margin of each completed survey. The writing must be neat and in a color different from that used by the person completing the survey. The coding work can be greatly reduced by preprinting the coding numbers and characters on the survey form. For our example, item 5 with preprinted code numbers would appear as:

5. How good a job did that doctor do in explaining your medical condition to you?
   (1)___ Excellent   (2)_X_ Very good   (3)___Good   (4)___Fair
   (5)___ Poor   (6)___ Don't remember

Dear Patient of Sayah Medical Clinic:

As part of their effort to monitor and improve the quality of their services, Sayah Medical Clinic has hired On Time Statistics to survey a sample of patients regarding their **most recent clinic visit**. It will take only about 5 minutes to complete this survey. On Time Statistics will provide a summary of the responses to Sayah Medical Clinic. Your responses will be completely confidential. Please help Sayah Medical Clinic by completing the survey. **Please mail the completed survey using the enclosed postage-paid envelope.**

Questions about this survey may be directed to: Karen Carson, On Time Statistics (Telephone 555-7828) or James Matthews, Sayah Medical Clinic (Telephone 555-2000).

1. Were you greeted by the receptionist when you entered the clinic?
   __X__ Yes _____ No _____ Don't remember
2. Did you meet with a nurse promptly at your appointment time?
   _____ Yes __X__ No _____ No appointment _____ Don't remember
   If no, how many minutes past your appointment time did you first meet with a nurse? __20__
3. Did you meet with a doctor?
   __X__ Yes _____ No _____ Don't remember
   If no or don't remember, please skip to question 8.
4. Who was the **first** doctor you met with on your **most recent visit** to Sayah Medical Clinic?
   _____ Dr. Alberts _____ Dr. Casey __X__ Dr. Jones _____ Dr. Sayah
   _____ Dr. Wesson _____ Don't remember doctor's name
5. How good a job did that doctor do in explaining your medical condition to you?
   _____ Excellent __X__ Very good _____ Good _____ Fair _____ Poor
   _____ Don't remember
6. How good a job did that doctor do in explaining the treatment plan to you?
   _____ Excellent __X__ Very good _____ Good _____ Fair _____ Poor
   _____ Don't remember
7. Did that doctor give you the opportunity to ask questions?
   __X__ Yes _____ No _____ Don't remember
8. Please rate the quality of service that you received from nurses on your **most recent visit.**
   _____ Excellent __X__ Very good _____ Good _____ Fair _____ Poor
   _____ Don't remember
9. How efficient was the cashier in processing your bill at the end of your **most recent visit?**
   _____ Excellent __X__ Very good _____ Good _____ Fair _____ Poor
   _____ Don't remember
10. Considering all aspects of your **most recent visit**, how would you rate the services you received?
    _____ Excellent __X__ Very good _____ Good _____ Fair _____ Poor

Thank you for helping Sayah Medical Clinic improve their services.

FIGURE 1.1 Completed survey

Another part of the data preparation process is the development of the data entry format. The nature of the format will depend on the software that will be used to analyze the data. For MINITAB®, a worksheet column is assigned to each survey item and to the survey identification number. For SAS®, an input statement is often created for data entry based on 80 columns per line. Specific columns are assigned for the responses to each survey item.

The data entry clerk (or you) then enters the coded data into a data file or worksheet using the data entry format. As with any human process, errors can and do occur in data entry. **It is important to check the data after data entry.**

If the data set is not too large, print it and check each entry against the corresponding coded survey. Entry-by-entry checks find many, but not all, errors. Computer routines that give sample sizes and five-number summaries or frequency tables for each variable are helpful in detecting data entry errors such as the entry of impossible values, misplaced decimal points, and items entered in the wrong columns. Look for unexpected results, such as sample sizes different from the number of completed surveys or unexpected values of a variable. Unexpected results are often caused by data entry errors. Scatter plots can also be helpful in detecting unusual entries. Correct all detected data entry errors before analyzing the data.

One strategy for detecting data entry errors is to enter the data twice using separate SAS data sets and using PROC COMPARE to detect all differences in the two data sets. This will detect all errors except those in which exactly the same data entry error was made for a particular item on a particular survey in both data sets. This method is very effective at finding data entry errors but doubles the data entry effort.

The following SAS code would detect a data entry error in the second observation for the variable b:

```
data one;
input a b c;
cards;
2 3 7
3 7 9
4 5 8
data two;
input a b c;
cards;
2 3 7
3 8 9
4 5 8
proc compare base = one compare = two;
run;
```

■ ■ ■ ■ ■ **PARTING GLANCES**

Sometimes it is difficult to know all data analyses, software packages, and computer platforms that will be used on a data set when the coding protocol and data entry format are developed. One person may look at the data using MINITAB on a Macintosh. Someone else may use SAS on a PC. A third person might use another package on a mainframe. It is wise to code and format the data so that it will be accessible in many different computing environments. Particular attention should be given to the treatment of missing data. Some packages treat a blank space as a missing value, whereas others treat it as a zero.

The information at the top of the survey form in Figure 1.1 is designed to encourage patient response. Notice that patients are told the purpose of the study, approximately how long it will take to complete the survey, that their responses will be confidential, and who to call if they have questions. They are also supplied a postage-paid envelope. This encourages response and ensures that surveys are mailed to the correct address.

The guarantee of confidentiality is an important obligation. When confidentiality is promised, you must ensure that it is kept. It is difficult to collect good data unless we are trusted by the public.

The Sayah Medical Clinic survey is a relatively small project with a single site. In such projects it is relatively easy for the statistician to compare the entered data with the coded survey forms. In larger studies, including those with multiple sites, the statistician often receives the data as a computer file and does not have immediate access to the data collection forms. In these cases, one cannot do item-by-item checks and must rely on the other error detection methods.

■ ■ ■ ■ ■ **ASSIGNMENT: PREPARING FOR DATA ANALYSIS**

1. Prepare a written coding protocol for the Sayah Medical Clinic survey. Your coding protocol should prescribe assignment of an identification number to each completed survey, prescribe codes for each expected response to each survey item, and give clear instructions regarding what to do if unexpected responses occur. The protocol should be written so that it is completely understandable to someone with a high school education.
2. Use your coding protocol to assign the survey identification number 001 and to code the completed survey in Figure 1.1. Remember to write neatly using a color different from black.
3. Prepare a written data entry format for entering the data using the software package of your choice. Indicate the software package that you will be using.

4. Suppose that the data from 300 completed surveys have been entered using your coding protocol and data entry format. Write a program for the software package of your choice (or explain all actions on a menu-based software package) that will do the following:
   a. Print the complete data set
   b. Allow you to detect the entry of a value that is not possible under the coding protocol
   c. Allow you to detect that all entries from a particular survey have been shifted by one column in the data entry (e.g., entered in columns 2–16 rather than 1–15)
   d. Produce a separate scatter plot of the responses to item 5 versus the responses to item 6 for each doctor in item 4
5. Explain the specific features of the computer output that you would look at to detect the types of errors mentioned in Problems 4b and 4c.

▪ ▪ ▪ ▪ ▪ ASSIGNMENT: ANALYZING THE DATA

Write a computer program (or explain all actions on a menu-based software package) that uses your coding protocol and data entry format to perform the following operations on the Sayah Medical Clinic data.

1. Construct frequency tables for the responses to each of the ten items.
2. Construct separate frequency tables for each staff member for items 2, 5, 6, 7, and 10.
3. Modify the data set so that the codes for "Don't remember" and "No response" are treated as missing values.
4. Compute the mean and median response to question 10 for each doctor.

CHAPTER 2

Designing a Telephone Survey

■ ■ ■ ■ ■ INTRODUCTION

The use of surveys to collect data is a major research tool in many fields, including the social sciences, marketing, education, and public health. In a survey study, individuals are asked to report information about themselves, their household, or their organization. As respondents report the data, it is crucial that all questions are understandable to them. To get the best possible information, the research team constructs and field-tests a carefully worded survey instrument and then uses it to gather information from a random sample of the population. Surveys are conducted by personal visits, over the telephone, and through traditional and electronic mail.

Computing Concepts and Procedures

Data entry computer screen

Mathematical Concepts

None

Statistical Concepts

Data recording form, survey instrument

Materials Required

None

■ ■ ■ ■ ■ **THE SETTING**

You are a statistician employed by Market Information Specialists Limited, a market research company. Bazoomi, Inc., produces an expensive encyclopedia product on CD-ROM for use with home computers. Bazoomi wishes to advertise the product on one of three television shows designed for family audiences, "Crazy Times in the Suburbs," "Those Were the Days," or "The Wally Walters Show." Your company has been hired by Bazoomi to conduct a telephone survey of a random sample of 1000 households in the United States to estimate the proportion of households in which each show is watched and to learn about the demographics of these households. Your boss, Stu Stewart, has assigned you the task of developing the survey script, which will be read over the telephone, and the data recording form. Other statisticians will handle sample selection, data entry, and data analysis.

Bazoomi's marketing director, Jacob Adman, believes the product is most likely to be purchased by households with the following characteristics: household income above $40,000, at least one adult with a college degree, a home computer, and at least one child between the ages of 6 and 18.

■ ■ ■ ■ ■ **THE BACKGROUND**

Telephone surveys differ from mail surveys in that interviewers need a script to read over the telephone. Having a well-written script reduces variation in how the questions are asked. The script for this project should accomplish the following:

1. Provide a friendly greeting to the person answering the phone.
2. Determine if the interviewer is speaking to an adult member of the household. If not, get an adult member of the household to the phone. If this is not possible, end the interview.
3. Provide the interviewer's first name and state that he or she is employed by Market Information Specialists Limited, a market research company.
4. Describe that the purpose of the call is to collect general marketing information for a major client and that no attempt will be made to sell anything.
5. Give the approximate amount of time that will be required to answer the questions.
6. State that all responses will be confidential.
7. Ask specific questions dealing with the household characteristics listed by Jacob Adman.

8. Ask specific questions to determine if anyone in the household watched the three television programs of interest to Bazoomi during the last two weeks.
9. Thank the person for their assistance.

In developing the script, you should encourage participation. One way to encourage participation is to minimize the length of the interview. Respondents are generally asked to volunteer their time and few people want to participate in a long telephone interview. We can minimize the length of the interview by not asking unnecessary questions and by eliminating needless words from the script.

Another way to encourage participation in a survey of the general adult population is to use language that is easily understood by people with an eighth-grade education. Don't confuse or intimidate the respondent.

A third way to encourage participation is to make the respondents comfortable with the data collection process. Giving a friendly greeting, telling the respondent the purpose of the survey, assuring confidentiality, and assuring that you aren't selling anything help make them comfortable. The form of the questions can make a difference to respondents. For personal questions such as age or income, many respondents may be much more willing to indicate an interval rather than give a precise value. For example, a 45-year-old man with an income of $71,230 per year might be willing to indicate that he is in the 40–49 age range and in the $50,000–$75,000 income range but unwilling to tell you his exact age or income.

As you develop the script, remember that the data must be coded. It is much easier to code responses to questions that have a small number of possible responses than questions with open-ended responses.

The person reading the script must have a method for efficiently recording the responses. This can be a data recording form or a customized data entry computer screen. The screen has the advantage of entering the data into the computer as it is being collected. Some survey researchers combine the script and the data recording form into a single document, either paper or electronic. If it is a paper document, one column is used for the script and another column is used for recording the data.

The form, screen, or combined script/form should indicate a sequential identification number for the respondent and have a simple method for quickly recording the responses to all questions. For example, if the ninth question in the script is "Does at least one adult in this household have a college degree?" the corresponding entry on a paper data recording form might be:

Q9 College Degree: (1) ___ Yes   (2) ___ No
      (9) ___ Unknown or No response

The numbers next to the responses are data entry codes.

■ ■ ■ ■ ■ **PARTING GLANCES**

The importance of having concise and easily understandable survey questions cannot be overemphasized. We cannot get good data if respondents don't understand a question or if a question has different meanings for different respondents. Producing good questions requires a great deal of effort. Alreck and Settle (1995), Converse and Presser (1986), and Fowler (1993) provide excellent advice on question construction.

It is crucial to field-test survey instruments before they are used to collect data from the random sample selected for a study. Research team members are often highly educated and have had a particular set of life experiences. Some respondents will have had very different educational and life experiences. Frequently, research teams construct questions that are perfectly logical from their viewpoint but are confusing or have multiple meanings to a respondent. Field-testing the instrument on individuals similar to the anticipated respondents will allow you to detect and correct most problems before the instrument is used to collect data from the random sample.

■ ■ ■ ■ ■ **REFERENCES**

Alreck, Pamela L., & Settle, Robert B. (1995). *The Survey Research Handbook: Guidelines and Strategies for Conducting a Survey* (2nd ed.). Homewood, IL: Irwin.

Converse, Jean M., & Presser, Stanley (1986). *Survey Questions: Handcrafting the Standardized Questionnaire*. Newbury Park, CA: Sage Publications.

Fowler, Floyd J., Jr. (1993). *Survey Research Methods* (2nd ed.). Newbury Park, CA: Sage Publications.

■ ■ ■ ■ ■ **ASSIGNMENT: PREPARING FOR DATA COLLECTION**

1. Develop a draft-written script for the Bazoomi telephone survey. Your script should accomplish the nine goals listed in The Background section. Use a style that encourages respondent participation.
2. Have a friend or classmate critique your script from the viewpoint of the person who will read the script to collect data and from the viewpoint of a respondent.
3. Prepare a data recording form to be used with your script. The form should indicate a sequential identification number for the respondent, have a simple method for quickly recording the responses to all questions, and indicate data entry codes, if applicable. If you prefer, you may combine the data recording form with your script.

4. Read your script to three friends over the telephone. Record their responses on your data recording form. Record the amount of time required to complete each interview.
5. Modify your script and data recording form, as necessary, based on what you learn in Problems 2 and 4.

C H A P T E R

3

Determining the Sample Size

■ ■ ■ ■ ■ INTRODUCTION

How large a sample do I need? Researchers frequently ask statisticians this question. Statisticians love data—the more the better. Larger sample sizes give more precise inferences. However, sample collection and processing require time and money, sometimes much time and money. Researchers, the ones who collect and pay for the data, like smaller sample sizes. Typically, the researcher and the statistician jointly determine the sample size. The statistician computes the power of a hypothesis test or the expected width of a confidence interval for different sample sizes. This allows the researcher to intelligently decide how much time and money to budget for the experiment. At times, the sample size corresponding to the maximum available resources will result in such low power or such wide confidence intervals that the study is not worth doing. The researcher must know this before the data collection. Otherwise, the researcher has wasted time and money.

Computing Concepts and Procedures

Noncentral t distribution function, numerical search, SAS function GAMMA, SAS function PROBT, SAS function TINV, Student's t distribution probability point

Mathematical Concepts

Gamma function, nonlinear inequality

Statistical Concepts

Confidence interval, expected value, function of a random variable, hypothesis test, power of a test, noncentral *t* distribution, Student's *t* distribution

Materials Needed

None

▪ ▪ ▪ ▪ ▪ THE SETTING

You are a statistician in the Statistical Consulting Group of XXY Manufacturing, Inc. Process engineer Don Gomez is investigating the mean amount of time required to complete an assembly task. He wishes to show that the mean assembly time is statistically significantly less than 30 seconds. He wants the hypothesis test to have a 0.05 significance level and to have at least a 90% chance of declaring the mean to be less than 30 seconds if the mean is actually 29.5 seconds or less. He also wishes to estimate the population mean using a 95% confidence interval with an expected width of at most 0.8 second. He has come to you for assistance in selecting the sample size. From experience with similar studies, he believes the assembly times will have an approximately normal distribution and that the sample range for 1000 observations would be approximately 9 seconds.

▪ ▪ ▪ ▪ ▪ THE BACKGROUND

Don Gomez is putting two constraints on the sample size. First, he needs a sample size large enough that the power of the hypothesis test is at least 0.90 if the actual mean is 29.5 seconds or less. Second, he needs a sample size large enough to make the expected width of the confidence interval at most 0.8 second. While these are reasonable requests from the engineer's viewpoint, the statistician needs more information before determining the necessary sample size. Let μ and σ denote the mean and standard deviation, respectively, of the distribution of assembly times.

Don Gomez wishes to test H_0: $\mu \geq 30$ seconds versus H_a: $\mu < 30$ seconds with a 0.05 significance level. Under the assumption that assembly times follow a normal distribution, the one-sample *t* test statistic is

$$T = \frac{\overline{X} - 30}{S/n^{1/2}} \tag{3.1}$$

The statistic has a Student's *t* distribution with $n - 1$ degrees of freedom if $\mu = 30$ seconds. Negative values of T suggest $\mu < 30$ seconds. Don should reject H_0 at the 0.05 significance level if $T \le -t_{n-1,0.05}$, the lower 0.05 point of the Student's t distribution with $n - 1$ degrees of freedom.

The power of the test is the probability of rejecting H_0, $P(T \le -t_{n-1,0.05})$. Under H_a, the test statistic T has a noncentral t distribution with $n - 1$ degrees of freedom and noncentrality parameter

$$\delta = \frac{\mu - 30}{\sigma / n^{1/2}} \tag{3.2}$$

Let $T_{n-1,\delta}$ denote a random variable having this noncentral t distribution. The power of this hypothesis test equals $P(T_{n-1,\delta} \le -t_{n-1,0.05})$. Thus, the power depends on μ, n, and σ. To satisfy Don's first goal of having n large enough so that the power is at least 0.90 if μ equals 29.5 seconds, the statistician must learn more about σ.

The statistician can often gain information about σ by interviewing the researcher. Researchers who are knowledgeable about statistics may be able to give you a very reasonable guess of σ based on their experience. If they are not as knowledgeable about statistics, it is often more productive to ask them for a reasonable guess of the minimum and maximum response they would expect in a sample of 1000 observations. For distributions close to the normal, the sample range for such a large sample is approximately 6σ. Thus, a guess of σ is the guess of the sample range divided by 6. If the researcher is more comfortable thinking in terms of 50 observations, a guess of σ is a guess of the sample range divided by 4. Another approach to gaining information about σ is to perform a pretest. That is, we collect a preliminary sample to estimate σ. Regardless of the approach, we use our estimate of σ in the sample size determination.

Once we have a preliminary guess of σ, we can determine the smallest sample size that will achieve our power goal. We are dealing with the nonlinear inequality in n,

$$P\left(T_{n-1\delta} \le -t_{n-1,0.05}\right) \ge 0.90 \tag{3.3}$$

Many statistical packages will compute the probability points and the distribution function of both the Student's t and noncentral t distributions. For example, if $n = 25$ and $\delta = -2.0$, the following SAS code computes $-t_{24,0.05}$ and $P\left(T_{24,-2.0} \le -t_{24,0.05}\right)$.

```
data one;
cp = tinv(0.05, 24);
power = probt(cp, 24, -2.0);
put cp power;
run;
```

We need to search over n to find the smallest sample size that will achieve the power goal. In performing the search, it helps to have a good initial guess of the solution. We can find an initial guess by replacing the *t* distributions by normal distributions. With this substitution, the inequality becomes

$$P\left(\frac{\overline{X}-30}{\sigma/n^{12}} \le -1.645\right) \ge 0.90 \tag{3.4}$$

After some manipulation, this new inequality becomes

$$P\left(\frac{\overline{X}-29.5}{\sigma/n^{1/2}} \le \frac{0.5-1.645\sigma/n^{1/2}}{\sigma/n^{1/2}}\right) \ge 0.90 \tag{3.5}$$

If $\mu = 29.5$, the left-hand side of the inequality is the standard normal distribution function evaluated at $-1.645 + 0.5n^{1/2}/\sigma$. The 90th percentile of the standard normal distribution is 1.282. Thus the new inequality is equivalent to

$$-1.645 + 0.5n^{1/2}/\sigma \ge 1.282 \text{ or } n \ge (5.854\sigma)^2 \tag{3.6}$$

Plugging in the preliminary guess of σ leads to the initial guess of *n*.

We evaluate $P(T_{n-1,\delta} \le -t_{n-1,0.05})$ for this initial guess of *n*. We then increase *n* until we find the smallest value of *n* that yields power of at least 0.90.

Under the assumption that assembly times follow a normal distribution,

$$\overline{X} \pm t_{n-1,0.025}S/n^{1/2} \tag{3.7}$$

is a 95% confidence interval for the population mean. The expected confidence interval width is $2t_{n-1,0.025}E(S)/n^{1/2}$. Although $E(S^2) = \sigma^2$, $E(S) \ne \sigma$. You will show in the section Assignment: Looking at the Mathematics that

$$E(S) = \frac{2^{1/2}\Gamma(n/2)\sigma}{(n-1)^{1/2}\Gamma[(n-1)/2]} \tag{3.8}$$

for a sample of size *n* from a normal distribution. Thus, the expected width of the 95% confidence interval is

$$2t_{n-1,0.025}\frac{2^{1/2}\Gamma(n/2)\sigma}{n^{1/2}(n-1)^{1/2}\Gamma[(n-1)/2]} \tag{3.9}$$

which depends on *n* and σ. Again, the statistician must learn more about σ to satisfy the process engineer's second goal of having *n* large enough that the expected confidence interval width is at most 0.8 second.

Once we have a preliminary guess of σ, we can find the smallest sample size to achieve the expected width goal. We have the nonlinear inequality in n,

$$2t_{n-1, 0.025} \frac{2^{1/2} \Gamma(n/2) \sigma}{n^{1/2} (n-1)^{1/2} \Gamma[(n-1)/2]} \le 0.8 \qquad (3.10)$$

Again, we need to search over n. The initial guess of n is obtained by simplifying the inequality to the corresponding form for known σ

$$2(1.96)\sigma/n^{1/2} \le 0.8 \qquad (3.11)$$

and solving for n. This yields an initial guess of n approximately equal to $(4.90\sigma)^2$.

We evaluate Equation (3.9) for this initial guess of n. We then increase n until we find the smallest value of n that will yield an expected confidence interval width of at most 0.8. The SAS function GAMMA can be used to evaluate the gamma function terms.

■ ■ ■ ■ ■ PARTING GLANCES

The required sample sizes depend on the preliminary estimate of σ. If the preliminary estimate is smaller than σ, the test will be less powerful and the expected width of the confidence intervals will be larger than anticipated. Conversely, if the preliminary estimate is larger than σ, the test will be more powerful and the expected width of the confidence intervals will be smaller than anticipated.

The initial guesses will be good when the required sample sizes are large. If the required sample sizes are small, the initial guesses may not be as good. The required sample sizes are usually larger than the initial guesses.

■ ■ ■ ■ ■ ASSIGNMENT: DETERMINING THE SAMPLE SIZE

1. Use the information in the section The Setting to determine a preliminary estimate of σ.
2. Use the preliminary estimate of σ to find the smallest sample size yielding power of at least 0.90 if $\mu = 29.5$ seconds.
3. Use the preliminary estimate of σ to find the smallest sample size that will yield an expected confidence interval width of at most 0.8 second.

▪ ▪ ▪ ▪ ▪ ASSIGNMENT: LOOKING AT THE MATHEMATICS

1. Let Y have a chi-square distribution with v degrees of freedom. Show that

$$E(Y^{1/2}) = 2^{1/2}\Gamma[(v+1)/2]/\Gamma(v/2)$$

   (Hint: Begin by writing the expected value as an integral and combining the powers of y.)

2. Using the results of Problem 1, show that

$$E(S) = \frac{2^{1/2}\Gamma(n/2)\sigma}{(n-1)^{1/2}\Gamma[(n-1)/2]}$$

   for a sample of size n from a normal distribution.

3. Let X and Y be independent random variables such that X has a normal distribution with mean θ and variance 1 and Y has a chi-square distribution with v degrees of freedom. Then $\dfrac{X}{(Y/v)^{1/2}}$ has a noncentral t distribution with v degrees of freedom and noncentrality parameter θ. Use this fact to show that for a sample of size n from a normal distribution with mean μ and variance σ^2, the statistic $T = \dfrac{\overline{X}-30}{S/n^{1/2}}$ has a noncentral t distribution with $n-1$ degrees of freedom and noncentrality parameter $\delta = \dfrac{\mu-30}{\sigma/n^{1/2}}$.

▪ ▪ ▪ ▪ ▪ ASSIGNMENT: THE WRITTEN PRESENTATION

Write a formal report to be read by process engineer Gomez and by your boss. Your report should include the following:

1. A brief introduction describing the experiment that Gomez wishes to perform and his goals for power and expected confidence interval width
2. A clear recommendation of the sample size that should be used
3. An appendix describing the effects of sample size on power and expected confidence interval width
4. Professional-quality tables and figures, as necessary

For guidance, refer to Chapter 12: Strategies for Effective Written Reports.

■ ■ ■ ■ ■ ASSIGNMENT: THE ORAL PRESENTATION

Prepare a 5-minute oral presentation, including professional-quality visual displays to present the results of your work, to Gomez, his boss, and your boss. Your presentation should address items 1–3 of The Written Presentation assignment. For guidance, refer to Chapter 13: Strategies for Effective Oral Presentations.

C H A P T E R

4

Designing an Experiment to Compare Two Horn Activation Buttons

■ ■ ■ ■ ■ **INTRODUCTION**

The research team must make several decisions regarding the design and analysis of an experiment. Consider the common problem of comparing two treatments. There are several characteristics that might be compared. For example, we might wish to compare the treatments with respect to a measure of central tendency, such as the mean or median; another measure of location, such as the 10th or 95th percentile; a measure of variability, such as the variance or interquartile range; or the probability that the treatment produces a specific response.

There are also experimental design choices. The experiment can be done as an independent sample design with subjects being randomly assigned to treatments. It can also be done as a dependent sample design. In the dependent sample design, subjects are paired such that they are as much alike as possible and then one member of each pair is randomly assigned to treatment one and the other to treatment two. In another dependent sample design approach, each subject is observed under both treatments. If this approach is used, randomization is used to decide whether the subject is observed first under treatment one or treatment two.

Given a characteristic to be compared and a design, there are often several statistical methods that might be used. In many cases, the choice of which method is best will depend on the data's underlying probability distribution. Some methods are based on specific distributional assumptions (e.g., normality) for the data. These assumptions may not be crucial if the sample sizes are large.

The statistician must talk with the other research team members to get the necessary information to make wise design and analysis decisions. It is a team effort.

Computing Concepts and Procedures

Random number generation, SAS function PROBT, two-dimensional plot

Mathematical Concepts

Derivative, inverse function

Statistical Concepts

Change of variable, density function, dependent sample design, independent sample design, normality assumption, power of a test, comparing two treatments

Materials Required

None

■ ■ ■ ■ ■ THE SETTING

You are a statistician in the Corporate Statistical Consulting Department of Modern Motors Corporation, an automobile manufacturer. Design engineer Latanya Walker and ergonomist Sam Carilli are considering a new design for the horn activation buttons on Modern Motors' most popular model. They want you to design an experiment and propose a data analysis to find out if drivers honk the horn faster using a new large oval button rather than using the current small square button. The relative sizes of the buttons and their positions on the steering wheel are shown in Figure 4.1. Latanya and Sam hope the new design will reduce the average time required to honk the horn by at least 5% for the population of all drivers.

■ ■ ■ ■ ■ THE BACKGROUND

Latanya Walker and Sam Carilli tell you that Modern Motors has groups of licensed drivers who meet regularly with personnel from several departments to discuss and evaluate various design and marketing strategies. They plan to use 20 drivers in an upcoming group to compare the two horn activation button designs.

They propose the following measurement protocol.

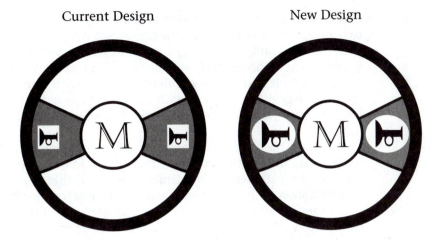

FIGURE 4.1 Current and new designs of horn activation button

1. The subject sits in the driver's seat of a car that has a steering wheel equipped with one of the two horn activation button designs.
2. The subject is shown how to adjust the seat and steering wheel positions and is allowed as much time as necessary to adjust them to his or her preferences.
3. The subject is shown the horn activation buttons on the steering wheel.
4. The subject is asked to grasp the steering wheel as he or she normally would while driving and then to honk the horn. The subject should repeat this step until he or she is comfortable honking the horn. If the subject has difficulty honking the horn, test personnel provide additional instruction, as necessary. If the subject is still unable to honk the horn, this fact is noted and the subject does not complete the rest of the measurement protocol.
5. The subject is shown a traffic light immediately in front of the car and is asked to tell the test personnel when the light changes from red to green. If the subject is unable to detect the light change in the first attempt, test personnel provide additional instruction, as necessary, and the process is repeated. If the subject is still unable to detect the light change, this fact is noted and the subject does not complete the rest of the measurement protocol.
6. The subject is again told to grasp the steering wheel as he or she normally would while driving and to honk the horn as soon as possible after the traffic light changes from red to green. The time from light change to horn activation is measured electronically to the nearest 1000th of a second. This step is repeated five times. The first two repetitions are considered practice. The average time for the last three repetitions is recorded as the subject's reaction time.

Latanya and Sam are thinking of randomly dividing the subjects into two groups of ten and measuring one group with the current horn activation button and the other group with the new button. If you think it is important to do so, they would be willing to consider having each subject complete the measurement protocol with one type of button and then repeat the protocol with the other type of button. This would involve more work for them and would require approval from their supervisors.

Based on his experience, Sam expects a large amount of subject-to-subject variation in the reaction times. It would not surprise him if one subject's measurement is four times larger than that of another subject. You ask Sam to draw a picture of what he thinks the distribution of reaction times for all drivers might look like with the current horn activation button. After some coaxing, Sam draws the density function shown in Figure 4.2.

When you ask him about within-subject variation, he explains that there are three steps involved. First, the subject must recognize that the light has turned green. Second, the subject must then move his or her hand to the vicinity of the button. Finally, the subject must push the button with enough force to honk the horn.

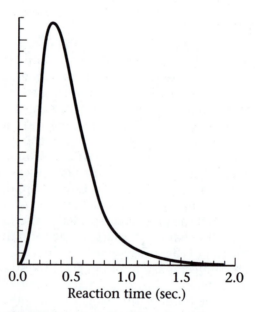

0.0   0.5   1.0   1.5   2.0
Reaction time (sec.)

FIGURE 4.2 Sam Carilli's guess of the density function of reaction time for all drivers with the current horn activation button design

Sam believes that the amount of time between recognizing that the light is green and touching in the vicinity of the horn activation button is relatively constant for repeated measurements on the same subject. However, there is more variation in the amount of time required to recognize that the light has turned green. He also tells you that a subject will occasionally miss the button when reaching for it or won't touch it with enough force to honk the horn. In such cases, the subject must adjust his or her hand before successfully honking the horn. This takes more time.

Overall, Sam thinks a subject's reaction time would be unlikely to change by more than 20% if they did the measurement protocol a second time using the same horn activation button design. He believes the density function of reaction times for a single subject would be skewed to the right but much less variable than the density function shown in Figure 4.2.

Latanya tells you that the new button's larger size is designed to make drivers less likely to miss the button. Drivers may also move more quickly to the vicinity of the new button because they may be more confident of hitting the larger button. She tells you that the amounts of force required to honk the horn with the current and new button designs are very similar.

You discuss this information with your boss, Wally Fisher. Wally suggests that you consider a model reflecting treatment effects, subject-to-subject variation, and within-subject variation. He tells you that modeling

$$\log(R) = (\text{treatment effect}) + (\text{subject effect}) + (\text{random error}) \qquad \textbf{(4.1)}$$

where R is reaction time, might be reasonable. Using $\log(R)$ rather than R should eliminate most of the skewness in Figure 4.2. If you observe the same subject under both treatments, the subject effect would be the same for the two observations, but the treatment effects and the random errors would be different. You will show in the section Assignment: Looking at the Mathematics that the density function of the difference of two independent, identically distributed random error terms is symmetric about zero.

Wally is also concerned that the sample sizes proposed by Latanya and Sam may make it difficult to detect the small differences in treatment effects that they hope to find. He suggests that you investigate the power of any test that you plan to do by using power tables, exact calculations, or computer simulation.

■ ■ ■ ■ ■ THINKING ABOUT THE MODEL

To evaluate the effectiveness of design and analysis plans, we need to translate the information that Latanya Walker and Sam Carilli have provided into knowledge about the terms in Equation 4.1. We will make a series of rough approximations to help you evaluate design and analysis plans.

Sam gave information about the density function of R for the current horn activation button in Figure 4.2. Denote this function by $f(r)$. The density of R peaks at about 0.35 second, and almost all of the density falls between 0.1 and 1.6 seconds.

Wally Fisher suggested that you model $W = \log(R)$ rather than R. The inverse function and its derivative are

$$R = \exp(W) \text{ and } \frac{dR}{dW} = \exp(W) \tag{4.2}$$

Using the change of variable technique, the density function of W is

$$g(w) = f[\exp(w)] \exp(w) \tag{4.3}$$

A graph of $g(w)$ would be approximately symmetric, with a peak at approximately –0.9 and most density falling between –2.3 and 0.5. A rough approximation is that W has a normal distribution with a mean of –0.9 and a standard deviation of

$$\frac{[0.5 - (-2.3)]}{6} = 0.467 \tag{4.4}$$

The variance of W is approximately $(0.467)^2 = 0.218$.

Now let's think about the right side of Equation 4.2. The treatment effect for the current horn activation button design, a constant, equals the mean of W, approximately –0.90. The subject effect and the random error terms are random variables with zero means. Let's assume, for now, that these variables have normal distributions.

Sam thought that a subject's reaction time would be unlikely to change by more than 20% if they did the measurement protocol a second time using the same horn activation button design. Increasing R by 20% increases $\log(R)$ by 0.182, and decreasing R by 20% decreases $\log(R)$ by 0.223. This would suggest that $W = \log(R)$ would be unlikely to change by more than approximately 0.20.

Differences in W measurements on the same subject and same button design are due to differences in the random error terms because both the treatment effect terms and the subject effect terms cancel. Let W_1 and W_2 denote the log reaction times for two observations on the same subject with the same button design. Let ε_1 and ε_2 denote the corresponding random error terms. Now, under the model in Equation 4.2,

$$P(|W_1 - W_2| < 0.20) = P(|\varepsilon_1 - \varepsilon_2| < 0.20) \tag{4.5}$$

Assuming that this probability is approximately .95, then 0.20 is approximately 1.96 standard errors of $\varepsilon_1 - \varepsilon_2$ or $1.96(2)^{1/2}$ standard deviations of ε_1. It follows that the standard deviation of the random error term is approximately 0.072. The variance of the random error term is approximately $(0.072)^2 = 0.005$.

If we make the usual assumption that subject effect and random error are independent, then the variance of the subject effect equals the variance of W minus the variance of the random error term: $0.218 - 0.005 = 0.213$.

Latanya and Sam hope the new design will reduce the average time required to honk the horn by at least 5%. Decreasing R by 5% decreases $W = \log(R)$ by 0.05. If Latanya and Sam are correct, the treatment effect for the new button design would be approximately $-0.90 - 0.05 = -0.95$.

Table 4.1 summarizes the results of our approximation. Although we can't assert that the results in Table 4.1 are correct, we hope that they are somewhat close to reality. We can use this information to evaluate different design and analysis plans in terms of their abilities to detect a 5% decrease in reaction times.

TABLE 4.1 Approximate Results for Use in Evaluating Design and Analysis Plans

Current button design effect	-0.90, constant
New button design effect	-0.95, constant
Subject effect	$N(0, 0.213)$, random
Random error	$N(0, 0.005)$, random

Latanya and Sam planned to use an independent sample design with ten observations per treatment. With this design, analysis would be done on the log reaction time values.

Under the assumed results in Table 4.1, the log reaction time values would have a $N(-0.90, 0.218)$ distribution for the current button design and a $N(-0.95, 0.218)$ distribution for the new button design. All observations would be independent.

The one-sided, two-sample t-test with a 5% significance level would declare the new button design to be better than the current button design if

$$T = \frac{\overline{W}_{new} - \overline{W}_{current}}{\left[S^2(1/10 + 1/10)\right]^{1/2}} < -1.734 \tag{4.6}$$

where S^2 is the pooled estimate of the variance. Under the assumed results of Table 4.1, T has a noncentral t distribution with $10 + 10 - 2 = 18$ degrees of freedom and noncentrality parameter

$$\delta = \frac{-0.95 - (-0.90)}{\left[0.218(1/10 + 1/10)\right]^{1/2}} = -.239 \tag{4.7}$$

Using power tables or the SAS function PROBT, we can find that the probability of declaring the new button to be better than the current button is .079. This suggests that we are very unlikely to detect a 5% improvement in reaction times using this design, sample sizes, and statistical method combination. We could also use computer simulation to approximate the power of nonparametric tests (such as the Wilcoxon rank sum) for this design. The power for these tests would also be very low for this design and sample size combination.

If we use a dependent sample design with all subjects observed under both treatments, the analysis would be done on the log reaction time for the new button design minus the log reaction time for the current button design differences. Subject effects cancel in these differences. Under the assumed results in Table 4.1, these differences would have a $N(-0.05, 0.010)$ distribution. The set of differences would be independent.

Under the assumptions of Table 4.1, the one-sample t-statistic has a noncentral t distribution with $n - 1$ degrees of freedom and noncentrality parameter $\delta = n^{1/2}(-0.05)/(0.010)^{1/2}$, where n is the number of subjects. The power of the one-sided t-test equals the probability that this noncentral t random variable is less than the lower α probability point of the t distribution with $n - 1$ degrees of freedom. The power can be found from power tables or from the SAS function PROBT. We could also use computer simulation to approximate the power of nonparametric tests, such as the sign test and the Wilcoxon signed-rank test, for this design.

Although the density function of W should be more nearly symmetric than the density function of R, there is no guarantee that the normality assumption is correct. You also may want to evaluate design and analysis plans using nonnormal distributions with the means and variances given in Table 4.1.

■ ■ ■ ■ ■ PARTING GLANCES

This chapter illustrates a relatively simple experiment to improve an automobile. The experiment has a single factor, button design, with two levels. Manufacturers use numerous planned experiments to improve their products. These experiments frequently involve several factors, each having two or more levels. Product improvements coming from these experiments help manufacturers remain competitive in world markets.

It is possible to compute the exact power of some hypothesis tests under certain assumptions. This was done for the independent sample t-test under the assumptions of Table 4.1. In many cases, we can't compute the power of a test. In those cases, statisticians estimate the power based on the results of a computer simulation.

For a computer simulation, statisticians have a computer generate a large number of random samples from specific probability distributions. Then they compute the test statistic for each sample. The proportion of these samples in which the null hypothesis would be rejected is an estimate of the power of the test.

Many computer packages will generate random variables from a wide variety of distributions. The computer code first generates uniform random variables and then transforms each variable to the desired distribution. For example, if U has the uniform distribution on $(0,1)$ and F is a continuous distribution function, then $F^{-1}(U)$ is a random variable with distribution function F. This is known as the probability integral transformation.

▪▪▪▪▪ ASSIGNMENT: DESIGN AND ANALYSIS DECISIONS

1. Do you believe the proposed plans for selecting subjects and collecting the measurements are free from bias? If not, how should they be improved?
2. Review the discussion in the section Thinking about the Model. Would you design the experiment as an independent sample or as a dependent sample design? Write instructions for the test personnel to use in implementing your design. These should include specific instructions on how subjects are assigned to treatments or the order of treatment presentation if subjects are to be observed under each treatment.
3. Based on the information you have been given, what statistical methods would you use with the data? What are the assumptions of these methods? Do you believe that these assumptions are likely to be satisfied with these data? If not, are the sample sizes large enough that this isn't a major concern?
4. Latanya and Sam are hoping for at least a 5% reduction in the average response over the population of all drivers. They plan to use 20 subjects. How likely are they to detect a 5% improvement using 20 subjects with your proposed design and analysis under the assumptions in Table 4.1? How many subjects do you recommend they use? Justify your answers through power calculations or computer simulations.

▪▪▪▪▪ ASSIGNMENT: LOOKING AT THE MATHEMATICS

1. Let X and Y be independent, identically distributed random variables. Show that the distribution of $X - Y$ is symmetric about zero.
2. Let X have a normal distribution with mean μ and variance σ^2. Find the density function of $Y = \exp(X)$. Find values of μ and σ^2 yielding a density function for Y similar to that shown in Figure 4.2.

■ ■ ■ ■ ■ **ASSIGNMENT: THE WRITTEN PRESENTATION**

Write a formal report to be read by design engineer Latanya Walker, ergonomist Sam Carilli, and your boss, Wally Fisher. Your report should include the following:

1. A brief description of the two horn activation buttons and the characteristic that Latanya and Sam wish to compare
2. A review of the sample selection procedure and the measurement protocol, including any suggestions that you might have to improve them
3. Your recommended experimental design, including sample sizes and instructions for the test personnel to use in implementing your design
4. Your recommendations for data analysis, including suggested statistics and probability points
5. Any assumptions of these statistical methods and the anticipated effects of assumption violations
6. Your estimate of the probability that your suggested analysis will be able to detect a 5% improvement in reaction time
7. Professional-quality tables and figures, as necessary

For guidance, refer to Chapter 12: Strategies for Effective Written Reports.

■ ■ ■ ■ ■ **ASSIGNMENT: THE ORAL PRESENTATION**

Prepare a 5-minute oral presentation, including professional-quality visual displays to present the results of your work, to design engineer Latanya Walker, ergonomist Sam Carilli, and your boss, Wally Fisher. Your presentation should address items 1–6 of The Written Presentation assignment. For guidance, refer to Chapter 13: Strategies for Effective Oral Presentations.

CHAPTER

5

Using Regression to Predict the Weight of Rocks

■ ■ ■ ■ ■ **INTRODUCTION**

Statisticians use models to summarize the relationships between two or more variables. One goal in regression modeling is to predict the value of the dependent variable from the values of one or more predictor variables. This is particularly valuable if the dependent variable is difficult or expensive to measure and the predictor variables are easy and cheap to measure.

In developing the regression model, the statistician chooses the predictor variable or variables, possibly transforms the variables, estimates the unknown parameters, and checks the validity of the model. At times, there is a natural predictor variable and a linear regression model is appropriate. In other situations, there are a large number of potential predictor variables and the statistician must determine which of these variables should be included in the model. The statistician must also determine if transformations are needed.

Computing Concepts and Procedures

SAS PROC PLOT, SAS PROC PRINT, SAS PROC REG, two-dimensional plot

Mathematical Concepts

Ellipsoid, rectangular solid

Statistical Concepts

Regression, residual, *r*-square, scatter plot

Materials Required

For each team, a caliper capable of measuring to the nearest 0.01 inch, a scale capable of measuring to the nearest gram, 20 rocks of various sizes but of similar composition, and muffin pans with holes numbered from 1 to 20. The rocks must have dimensions and weights within caliper and scale capacities.

■ ■ ■ ■ ■ THE SETTING

You are a statistician working as a private consultant to the Greendale School District. Your job is to help classroom teachers develop hands-on science experiments that illustrate uses of mathematics and statistics. High school science teacher Nancy Stone asks you to develop a regression model for predicting the weight of small rocks from their external dimensions. The available measurement devices are a caliper and a scale.

■ ■ ■ ■ ■ THE BACKGROUND

Rocks are irregularly shaped, three-dimensional objects. A rock's weight depends on its volume, its chemical composition, and possibly other factors. The available measurement devices cannot directly measure volume, chemical composition, or unknown factors.

You can approximate a rock's volume from its external dimensions. For example, you can think of a rock as being approximately a rectangular solid. If you develop operational definitions of the height, width, and depth of a rock, then the product of these variables is the volume of the rectangular solid that approximates the rock. We will refer to this approximation of a rock's volume as its rectangular solid volume.

You might expect rock weight to be a constant × volume + random error. This gives a no-intercept simple linear regression model. Because rectangular solid volume approximates true volume, we might hope that the relationship between rock weight and rectangular solid volume will be linear. It is unclear if an intercept is needed. It is better to initially include an unneeded intercept than to leave out a needed one. The *t*-test for testing that the intercept is zero can help you decide whether to leave the intercept in the model.

Rock width is related to true volume, but it is not an approximation of it. Multiplying width by height and depth yields an approximation of volume. Because the logarithm of a product is the sum of the logarithms of the factors, log(weight) might be a linear function of log(width). If so, an intercept would probably be needed.

Nancy Stone would like you to develop two regression models. The first model will predict rock weight, using rock width as the predictor variable. The second model will predict rock weight, using rectangular solid volume. After you develop these models, she will want your advice regarding which model to have her students use.

One method of assessing the goodness of a regression model is its *r*-square value. If a model has an intercept, then the *r*-square value is the proportion of variability in the dependent variable that is explained by the model. Many computer packages multiply this proportion by 100 to make it a percentage.

Some computer packages will give an *r*-square value for no-intercept regression models. The *r*-square value for these models is the proportion, or percentage, of variability in the dependent variable *from zero* that is explained by the model. This value is not as useful as the *r*-square value given for models with intercepts.

The *r*-square values for models with intercepts are not comparable to *r*-square values for models without intercepts. It is better to compare error mean squares for the models or other measures of fit. The *r*-square values are also not comparable for a model that transforms the dependent variable with one that doesn't, or for models that use different transformations of the dependent variable.

Adding an additional predictor variable to a model always increases the *r*-square value or leaves it unchanged. Thus, comparing the *r*-square of a model to the *r*-square of the model with additional predictor variables can be misleading. Many computer packages include an adjusted *r*-square. This value adjusts the *r*-square value for the number of predictor variables in the model. Adding an additional predictor variable to a model increases the adjusted *r*-square if and only if the error mean square with the additional variable is less than the error mean square without the additional variable. You can also compare the error mean squares for the models or other measures of fit.

Because the relationship between weight and volume is different for rocks with different chemical compositions, different regression models would be necessary for different types of rocks. To keep things simple, you might suggest that the students use rocks with similar chemical compositions.

■ ■ ■ ■ ■ THE EXPERIMENT

Step 1: Operational Definitions

Before collecting data, we need to carefully define all variables. These written definitions must state exactly how the measurements will be made, including the unit of measurement. The definitions must be

TABLE 5.1 Operational Definitions of Rock Variables

Weight is the weight of a rock (to the nearest gram) measured with a scale.
Width is the maximum external dimension of a rock (to the nearest 0.01 inch) measured with a caliper.
Height is
Depth is
Rectangular solid volume = width × height × depth

understandable to high school students. Precise operational definitions reduce confusion and measurement error.

Operational definitions for weight and width are given in Table 5.1. You need to write operational definitions for height and depth. **Read this entire step before writing these definitions.**

The definitions of width, height, and depth should reflect the fact that these dimensions are perpendicular to each other. The orientation of a rock (the front, back, top, bottom, etc.) is arbitrary. In developing operational definitions, be sure that the rectangular solid volume does not change if the rock is turned.

Some variables are hard to describe. Considerable editing of your definitions may be necessary. Write your first drafts on a separate sheet. When you think you have good definitions, have other students or your instructor review them. A good way to find flaws in operational definitions is to observe others trying to make the measurement without any instruction other than your definition. If they are not making the measurement correctly, your definition may need more work. After your operational definitions have been revised, as necessary, write them in Table 5.1.

TABLE 5.2 Rock Weights and Dimensions

ROCK	WEIGHT	WIDTH	HEIGHT	DEPTH
1				
2				
3				
4				
5				
6				
7				
8				
9				
10				
11				
12				
13				
14				
15				
16				
17				
18				
19				
20				

Step 2: Data Collection

Place the 20 rocks in the numbered muffin pan holes. The experiment consists of measuring the variables weight, width, height, and depth of each rock. Use the operational definitions in making the measurements. Write the data in Table 5.2. Return each rock to its muffin pan hole after making the measurements. This will allow you to locate the rock if you have to check the measurements.

Step 3: Data Editing

Enter the data into a statistical computing package, and then double check the data entry. When you are sure the data have been correctly entered, create the rectangular solid volume, log(weight), and log(width) variables.

You are now ready to investigate the relationships between weight and the predictor variables, width and rectangular solid volume. Begin this investigation by creating scatter plots for the following pairs of variables:

weight and width
log(weight) and log(width)
weight and rectangular solid volume

Look for outliers in the scatter plots. These are points separated from the data cloud. Outliers are caused by data entry errors, by measurement errors, and by rocks that are very different from the other rocks in your sample. If you find outliers, determine the rock numbers of the unusual points. Then, determine the cause of each outlier by checking the data entered for the rock, remeasuring the rock, and visually comparing the rock with the other rocks. If data entry or measurement errors have been made, correct them and make new scatter plots. With measurement errors, also correct Table 5.2.

Step 4: Fitting the Models and Assessing the Fit

After editing the data, look for patterns in the scatter plots. The Background section suggests that the last two scatter plots might show linear patterns. After looking at the patterns, fit two separate regression models for weight or log(weight), using the predictor variables width and rectangular solid volume, respectively. Transform the predictor variables, as needed. It is your decision whether to include an intercept.

A residual is the difference between the observed value of the dependent variable and the value of the dependent variable predicted by the model. After fitting the regression models, create scatter plots of the residuals versus the predictor variable for both of your models. If a residual scatter plot exhibits curvature, then a transformation of the predictor variable or a higher-order term may be needed. With no-intercept models, an upward or downward trend in the residuals versus the predictor variable suggests an intercept is needed.

After examining the residual plot for each model, you must either accept the model or fit a new model. If you accept the model, summarize it in Table 5.3. Otherwise, fit the new model and examine the residual plot. Continue this process until you accept a model, and then summarize it in Table 5.3.

The following SAS code enters the data, creates the additional variables, lists the data set, and produces the scatter plots:

TABLE 5.3 Summary of Two Regression Models

Model involving width (include regression coefficient estimates)			
_____ = _____			
SOURCE	**DF**	**SS**	**MS**
Model	____	_____	_____
Error	____	_____	_____
Total	____	_____	_____
r-square = _____		Adjusted r-square = _____	

Model involving rectangular solid volume (include regression coefficient estimates)			
_____ = _____			
SOURCE	**DF**	**SS**	**MS**
Model	____	_____	_____
Error	____	_____	_____
Total	____	_____	_____
r-square = _____		Adjusted r-square = _____	

```
data one;
input weight width height depth;
rsv = width * height * depth;
logwt = log(weight);
logwidth = log(width);
cards;
28 1.68 1.09 0.82
                        An entry for each rock goes here.
12 1.07 1.02 0.63
proc print;
proc plot;plot weight*width logwt*logwidth weight*rsv;
run;
```

The following additional SAS code fits the regression lines

weight = α(rectangular solid volume) + error

log(weight) = β_1 + β_2[log(width)] + error

$$(5.1)$$

and makes the residual plots:

```
proc reg;
model weight = rsv / noint r;
output out = two residual = resid1;
proc reg;
model logwt = logwidth / r;
output out = three residual = resid2;
proc plot;plot resid1*rsv resid2*logwidth;
```

These models are included to illustrate SAS code and are not necessarily the best models for your data.

■ ■ ■ ■ ■ PARTING GLANCES

The use of calipers to approximate the volume of such irregular shaped objects is extremely crude. A much better approximation of the rock's volume can be made by filling a container with water, gently placing the rock in the water, and measuring the volume of the displaced water. The water displacement method works well with small rocks, but it would be more difficult with very large rocks.

An alternative to approximating a rock by a rectangular solid is to approximate it by an ellipsoid. We did not consider ellipsoids because their volumes are a constant times the corresponding rectangular solid's volume. It follows that models based on rectangular solids and on ellipsoids lead to the same predictions of weight.

■ ■ ■ ■ ■ ASSIGNMENT: ANALYZING THE DATA

Include copies of Tables 5.1, 5.2, 5.3, and the residual plots for the models in Table 5.3 with your responses.

1. Did you detect any outliers in your initial scatter plots? If so, explain how you detected that an outlier was present. Was this observation an outlier because someone made a data entry error, because someone made a measurement error, or because the rock was very different from the other rocks? How did you come to this conclusion about the cause of the outlier? What changes, if any, were made to your data set after you detected the outlier?
2. For both models in Table 5.3, produce a plot showing your original data and the fitted regression model.
3. In your opinion, which model gives the better fit of the data? Explain your answer.
4. Which model do you recommend that Nancy Stone use with her high school science students? Explain your answer.

5. What, if anything, would you do differently if you were to redo this experiment?

■ ■ ■ ■ ■ ASSIGNMENT: LOOKING AT THE MATHEMATICS

1. What is the volume of an ellipsoid having axes of lengths w, h, and d?
2. Argue that a regression model using ellipsoid volume as the predictor variable and the regression model using rectangular solid volume as the predictor variable would give the same values of predicted weight.
3. Would the regression coefficients be the same for the models using either rectangular solid volume or ellipsoid volume as the predictor variables? Explain your answer.

■ ■ ■ ■ ■ ASSIGNMENT: THE WRITTEN PRESENTATION

In your role as a private consultant to the Greendale School District, write a report to be read by high school science teacher Nancy Stone and the district science coordinator summarizing your work. Your report should include the following:

1. A summary of the experiment you performed and its purpose
2. The operational definitions of all variables
3. A description of the measurement devices
4. A description of your two regression models and any assumptions that were made
5. Assessments of the goodness of fit of the models
6. A clear recommendation of the model that Nancy Stone's high school students should use in their hands-on science experiment
7. Any other recommendations you wish to make regarding using the experiment with high school students
8. An appendix containing the raw data
9. Professional-quality tables and figures, as necessary

For guidance, refer to Chapter 12: Strategies for Effective Written Reports.

■ ■ ■ ■ ■ ASSIGNMENT: THE ORAL PRESENTATION

In your role as a private consultant to the Greendale School District, prepare a 5-minute oral presentation, including professional-quality visual displays, to present the results of your experiment to Nancy Stone and the district science coordinator. Your presentation should address items 1–7 and 9 of The Written Report assignment section. For guidance, refer to Chapter 13: Strategies for Effective Oral Presentations.

CHAPTER

6

Estimating Variance Components in Tack Measurements

■ ■ ■ ■ ■ **INTRODUCTION**

There is variability in items produced by any manufacturing process. "Identical"-looking items are not identical. They differ in length, width, weight, and so on. A central goal in quality improvement is to reduce variation. Operators routinely measure a sample of items from a production process to monitor the variation. Unfortunately, the measurement process is also subject to variation. The measurement of an item's length is its actual length plus a measurement error. Repeated measurements on the same item may give different results. The variation in these measurements may increase if the measurements are made by different operators or with different measurement devices.

Measurement errors cloud the manufacturer's view of the actual variation among mass-produced items. We wish to reduce measurement errors as much as possible to more clearly understand the actual variation among the items. If measurement errors are negligible, we can have a high degree of confidence in our measurements. In this chapter we will study sources of measurement error and compare the measurement variation to the actual variation in carpet tack lengths.

Computing Concepts and Procedures

SAS PROC VARCOMP

Mathematical Concepts

System of linear equations

Statistical Concepts

Analysis of variance, factorial design, nested design, random effect, replicate, unbiased estimator, variance component

Materials Required

For each team, four nominal ½-inch carpet tacks, three micrometers capable of measuring to 0.001 inch, two objects with a premeasured dimension of less than 1 inch, 12-inch length of masking tape

Measuring with a Micrometer

Figure 6.1 identifies the parts of a micrometer. Place the item to be measured between the anvil and spindle, and close the micrometer by rotating the clutch clockwise until the thimble no longer rotates. **Warning:** Close the micrometer by rotating the **clutch** and not the **thimble**. Closing by rotating the thimble can damage the micrometer and the object being measured.

You determine the anvil-to-spindle distance by observing the markings on the barrel and the thimble. On the barrel, major divisions represent units of 0.100 inch and finer divisions represent units of 0.025 inch. The thimble markings represent units of 0.001 inch. The measurement depicted in Figure 6.1 represents two major divisions (2 × 0.100″ = 0.200″), three finer divisions (3 × 0.025″ = 0.075″), and seven units on the thimble (7 × 0.001″ = 0.007″), for a total measurement of 0.200 + 0.075 + 0.007 = 0.282 inch.

Your instructor may provide some objects with known dimensions. If so, practice measuring them before you begin the experiment. If your measurement differs from the known dimension by more than 0.002 inch, request assistance.

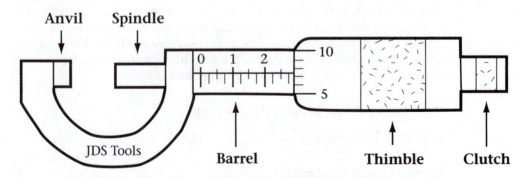

FIGURE 6.1   Micrometer at 0.282-inch setting

■ ■ ■ ■ ■   **THE SETTING**

You are a statistician working in the Quality Department of the Sharp Point Tack Company. The company has recently set up a new factory. Plant manager William Bossman wants an evaluation of the process for measuring the length of nominal ½-inch carpet tacks. Operators routinely take these measurements as part of the plant's quality-control effort. The measurement device is a micrometer.

■ ■ ■ ■ ■   **THE BACKGROUND**

A tack length measurement equals the true length plus measurement error. The true tack lengths vary because of variation in the manufacturing process and in the raw materials. There are three sources of measurement error.

The first source of measurement error is operator-to-operator differences. Operators differ in their technique for placing tacks in the micrometer, in the amount of force they apply when closing the micrometer, and in reading the measurement. These differences cause some operators to make measurements that are, on average, higher than those of other operators. Operator-to-operator differences can be reduced by training the operators to follow an easily understood measurement protocol.

The second source of measurement error is micrometer-to-micrometer differences. Micrometers must be calibrated frequently. If a micrometer is not calibrated correctly, it may tend to overestimate lengths (positive bias) or underestimate lengths (negative bias). Micrometer-to-micrometer differences can be reduced by frequent calibration and by the replacement of defective micrometers.

The third source of measurement error is inherent variability (lack of repeatability) in the measurement process. If a worker makes repeated length measurements on a single tack with the same micrometer, the measurements will vary. This inherent variability is independent of operator-to-operator and micrometer-to-micrometer differences. Inherent variability reflects small differences in the length measurement related to the part of the tack head touching the micrometer and in the snugness of the tack in the micrometer when the measurement is made. Reduction in the inherent variability generally requires a fundamental change in the measurement process. Replacing the micrometer with a laser-based measurement device would represent such a fundamental change.

Notice that reducing the different sources of measurement error requires different corrective actions: training of operators, calibration of micrometers, and capital investment for improved measurement devices. We will design an experiment to estimate the amount of variation due to each source. These estimates will show the relative importance of each source and give guidance regarding any necessary corrective action to the measurement process. We also wish to estimate the amount of tack-to-tack variation present in the production process.

■ ■ ■ ■ ■ **THE EXPERIMENT**

Step 1: The Basic Measurement

Our basic measurement is having an operator measure the length of a nominal ½-inch carpet tack to the nearest 0.001 inch with a micrometer. Tack length is defined as the distance from the head of the tack to the end of the point. Figure 6.2 illustrates the length.

Length

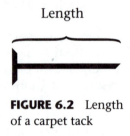

FIGURE 6.2 Length of a carpet tack

Step 2: Data Collection

Use masking tape to clearly label the micrometers as 1, 2, and 3 and the tacks as 1, . . . , 4. Three team members will be identified as operators 1, 2, and 3. It is crucial that we can identify specific micrometers, operators, and tacks.

There will be 72 measurements in this experiment. Each of the three operators will carefully measure the length of the four tacks twice with each of the three micrometers. The repeated measurements of a tack by the same operator with the same micrometer are known as replicates (or reps). We are using a three-factor factorial design because each operator is measuring each tack with each micrometer.

Your student intern, Wannabe A. Statman, has generated a random order for these 72 measurements. This random order appears in Table 6.1. The random order is used to reflect possible changes in experimental conditions during the course of the experiment. Examples of such changes include temperature changes, which cause the tacks to expand or contract, and compression of the tacks in the measurement process.

The operators should now make the 72 measurements, following the order given in Table 6.1. Make sure that the correct operator, micrometer, and tack are used for each measurement. **Don't cheat!** It is important to the success of the study that all 72 measurements be made independently;

TABLE 6.1 Random Order of Micrometers (M), Operators (O), and Tacks (T)

ORDER	M	O	T	√	ORDER	M	O	T	√	ORDER	M	O	T	√
1	2	2	2		25	2	2	3		49	3	3	1	
2	1	1	4		26	3	1	2		50	2	3	1	
3	2	2	3		27	3	2	1		51	3	3	2	
4	1	2	2		28	1	3	3		52	1	3	1	
5	1	2	3		29	2	2	4		53	2	1	3	
6	1	3	1		30	3	2	1		54	2	1	1	
7	2	3	2		31	2	2	1		55	1	3	4	
8	3	3	4		32	1	2	3		56	1	2	1	
9	3	2	4		33	3	2	3		57	1	1	1	
10	2	3	2		34	1	2	2		58	2	3	3	
11	2	2	2		35	1	1	3		59	1	3	3	
12	1	3	2		36	2	2	1		60	2	1	1	
13	2	1	4		37	1	1	4		61	1	3	2	
14	2	1	2		38	3	1	1		62	2	3	1	
15	3	1	4		39	2	1	4		63	1	3	4	
16	3	3	3		40	3	3	1		64	3	1	4	
17	3	1	1		41	3	3	2		65	2	3	4	
18	1	1	3		42	3	3	3		66	1	1	2	
19	3	1	2		43	3	2	3		67	1	1	2	
20	3	2	4		44	2	2	4		68	1	2	1	
21	2	1	2		45	2	3	3		69	3	1	3	
22	1	1	1		46	3	2	2		70	1	2	4	
23	2	1	3		47	3	3	4		71	1	2	4	
24	3	2	2		48	2	3	4		72	3	1	3	

that is, without referring to the other measurements. After a measurement is taken, enter a check mark (√) in Table 6.1, write the measurement in the appropriate space in Table 6.2, and return the tack and micrometer to the work area.

TABLE 6.2 Tack Length Measurements

MICROMETER	OPERATOR	REPLICATE	TACK 1	TACK 2	TACK 3	TACK 4
1	1	1				
		2				
	2	1				
		2				
	3	1				
		2				
2	1	1				
		2				
	2	1				
		2				
	3	1				
		2				
3	1	1				
		2				
	2	1				
		2				
	3	1				
		2				

Step 3: The Random Effects Model

Throughout this session, assume that the operators, micrometers, and tacks have already been selected as random samples from their respective large populations. Our interest is in variation within these populations and is not restricted to the individual operators, micrometers, and tacks used in the experiment. Developing the sampling plan would be part of the statistician's job in the factory.

Let Y_{ijkm} denote length measurement m of tack i by operator j using micrometer k for $i = 1, \ldots, 4; j = 1, 2, 3; k = 1, 2, 3;$ and $m = 1, 2$. As the operators, micrometers, and tacks are assumed to have been randomly selected, we are dealing with the three-factor, random effects model

$$Y_{ijkm} = \mu + T_i + O_j + M_k + \varepsilon_{ijkm} \tag{6.1}$$

where μ is an overall mean, T_i is a tack effect random variable having variance σ_T^2, O_j is an operator effect random variable having variance σ_O^2, M_k is a micrometer effect random variable having variance σ_M^2, and ε is a random error term having variance σ_ε^2. The parameters σ_T^2, σ_O^2, and σ_M^2 reflect the amounts of tack-to-tack, operator-to-operator, and micrometer-to-micrometer variation, respectively. The parameter σ_ε^2 reflects the amount of inherent variability. Large values of these parameters suggest a large amount of variation. For simplicity, interaction terms have not been placed in the model. Tack-to-tack, operator-to-operator, and micrometer-to-micrometer differences can be expected to be independent in this experiment.

We need to estimate the variances σ_T^2, σ_O^2, σ_M^2, and σ_ε^2. These parameters are known as variance components because the variance of the random variable Y_{ijkm} equals the sum of these four parameters. Estimates of the variance components are functions of the analysis of variance mean squares. The analysis of variance table for the model is shown in Table 6.3. The "dot" notation in the terms in the sums of squares reflects averaging over the values of the replaced subscript. Thus, $\bar{Y}_{i...}$ denotes the average of the 18 length measurements for tack i, $\bar{Y}_{.j..}$ is the average of the 24 measurement made by operator j, and so on.

The expected values of the mean squares are linear functions of the variance components. Setting the observed mean squares equal to their expected values and solving for the variance components yields the following unbiased estimators for the components:

TABLE 6.3 Analysis of Variance Table for Model in Equation (6.1)

SOURCE	D.F.	SUM OF SQUARES	MEAN SQUARE	EXPECTED MEAN SQUARE
Tack	3	$18\sum_{i=1}^{4}\left(\bar{Y}_{i...}-\bar{Y}_{....}\right)^2$	$\dfrac{SS_{tack}}{3}$	$18\sigma_T^2 + \sigma_\varepsilon^2$
Operator	2	$24\sum_{j=1}^{3}\left(\bar{Y}_{.j..}-\bar{Y}_{....}\right)^2$	$\dfrac{SS_{operator}}{2}$	$24\sigma_O^2 + \sigma_\varepsilon^2$
Micrometer	2	$24\sum_{k=1}^{3}\left(\bar{Y}_{..k.}-\bar{Y}_{....}\right)^2$	$\dfrac{SS_{micrometer}}{2}$	$24\sigma_M^2 + \sigma_\varepsilon^2$
Error	64	$SS_{total} - SS_{tack} - SS_{operator}$ $- SS_{micrometer}$	$\dfrac{SS_{error}}{64}$	σ_ε^2
Total	71	$\sum_{i=1}^{4}\sum_{j=1}^{3}\sum_{k=1}^{3}\sum_{m=1}^{2}\left(\bar{Y}_{ijkm}-\bar{Y}_{....}\right)^2$		

$$\hat{\sigma}_T^{\ 2} = \frac{MS_{tack} - MS_{error}}{18} \qquad \hat{\sigma}_O^{\ 2} = \frac{MS_{operator} - MS_{error}}{24}$$

(6.2)

$$\hat{\sigma}_M^{\ 2} = \frac{MS_{micrometer} - MS_{error}}{24} \qquad \hat{\sigma}_\varepsilon^{\ 2} = MS_{error}$$

An example trial of this experiment yielded the following mean-square values:

$$MS_{tack} = 0.00266228 \qquad MS_{operator} = 0.00003633$$

(6.3)

$$MS_{micrometer} = 0.00001550 \qquad MS_{error} = 0.00000268$$

Your mean-square values will be different because you will use different tacks, operators, and micrometers. Estimates of the variance components for the example data and estimates of the proportion of variation due to each component are given in Table 6.4.

For the example data, almost all variation is due to tack-to-tack differences. Also, inherent variability appears to be a bigger source of measurement error than either operator-to-operator or micrometer-to-micrometer differences.

Although variance component estimators are unbiased, they occasionally produce negative estimates of small variance components. Because all variance components are nonnegative, it is common to replace negative estimates with zero.

TABLE 6.4 Estimates of Variance Components and Percentages of Total Variation for Example Data

VARIANCE COMPONENT	ESTIMATE	% OF TOTAL VARIATION
σ_T^2	$\dfrac{0.00266228 - 0.00000268}{18} = 0.00014776$	96.97
σ_O^2	$\dfrac{0.00003633 - 0.00000268}{24} = 0.00000140$	0.92
σ_M^2	$\dfrac{0.00001550 - 0.00000268}{24} = 0.00000053$	0.35
σ_ε^2	0.00000268	1.76
Total	0.00015237	100.00

Many statistical computing packages will estimate variance components. The following SAS code analyzes the example data:

```
data one;
input operator microm tack length;
cards;
1 1 1 0.548
          An entry for each observation goes here.
3 3 6 0.563
proc print;
proc varcomp method=type1;
class operator microm tack;
model length=operator microm tack;
run;
```

This SAS code uses a separate line for entering each of the 72 observations. Alternative formats allow for multiple observations per line.

■ ■ ■ ■ ■ **PARTING GLANCES**

Making the 72 measurements in the random order specified by Table 6.1 is cumbersome. It would be much simpler for an operator to make both measurements on the same tack with the same micrometer at the same time. However, taking both measurements at the same time would violate our assumption that the random error terms are independent. The results of the first measurement could influence the second measurement.

We used a replicated, three-factor, random effects, factorial design to estimate the variance components. With this design, each of the three operators measured each of the four tacks twice with each of the three micrometers. This is not the only possible design for this study. One alternative approach is a three-factor nested design. An example of a nested design would use three operators, nine micrometers, and thirty-six tacks. Each operator would be issued three bags. Each bag would contain a randomly selected micrometer and four randomly selected tacks. Each tack would then be measured twice with the micrometer from the same bag. Figure 6.3 illustrates this nested design but does not indicate that each tack is measured twice.

This nested design requires the same number of observations as the factorial design, but the allocation of degrees of freedom is quite different. The degrees of freedom for the factorial and nested designs are shown in Table 6.5. The factorial design allocates more degrees of freedom to error, whereas the nested design allocates more degrees of freedom to tacks and micrometers. The two designs also use different models and different sum of squares formulas.

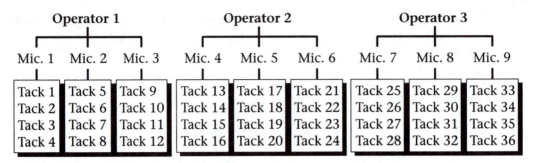

FIGURE 6.3   Three-factor nested design

We estimated the variance components using functions of the mean squares. Some quality-improvement practitioners prefer to estimate variability using the sample range. Although the range is easier to compute, it does not provide as much information as methods based on mean squares (unless only two measurements are used to compute the range and mean square). We want to get as much information as possible from our data, and computer packages make it relatively easy to find the variance component estimates.

TABLE 6.5   Degrees of Freedom for Factorial and Nested Designs

SOURCE	DEGREES OF FREEDOM FACTORIAL DESIGN	DEGREES OF FREEDOM NESTED DESIGN
Tack	3	27
Operator	2	2
Micrometer	2	6
Error	64	36
Total	71	71

■ ■ ■ ■ ■ ASSIGNMENT: ANALYZING THE DATA

Include a copy of Table 6.2 with your responses.

1. Use a statistical computing package to list your data and to compute the mean squares for tacks, operators, micrometers, and error. Make sure the printed data matches the data in Table 6.2.
2. Using either a statistical computing package or manually, compute estimates of the four variance components.

3. Comment on the relative sizes of the variance component estimates. What actions should the plant manager consider, based on your results? Explain your answer.

■ ■ ■ ■ ■ ASSIGNMENT: LOOKING AT THE MATHEMATICS

1. Substitute the model for Y_{ijkm} given in Equation (6.1) into the expression for MS_{tack} given in Table 6.3. Use the laws of expectation to prove that $E(MS_{tack}) = 18\sigma_T^2 + \sigma_\varepsilon^2$. Assume that all random variables are independent and have means equal to zero.

■ ■ ■ ■ ■ ASSIGNMENT: THE WRITTEN PRESENTATION

Turn in a copy of Table 6.2 separate from your written presentation.

In your role as a quality-improvement statistician, write a formal report summarizing your work to be read by plant manager William Bossman and Quality Department head Q. C. Gauge. Your report should include the following:

1. A summary of the experiment that was performed and its purpose
2. A description of the measurement protocol and the operational definition of tack length
3. A description of the populations from which the tacks, operators, and micrometers were selected
4. Potential sources of variation in the measurements
5. The random effects model, including descriptions of the variance components
6. The experimental design
7. The analysis of variance table and estimates of the variance components
8. Interpretation of the statistical results and recommended actions to be considered based on these results
9. Suggestions for future experiments or improvements to this experiment
10. Professional-quality tables and figures, as necessary

For guidance, refer to Chapter 12: Strategies for Effective Written Reports.

■ ■ ■ ■ ■ ASSIGNMENT: THE ORAL PRESENTATION

In your role as a quality-improvement statistician, prepare a 5-minute oral presentation, including professional-quality visual displays, to present the results of this experiment to plant manager William Bossman and Quality Department head Q. C. Gauge. Your presentation should address items 1–10 of The Written Presentation assignment section. For guidance, refer to Chapter 13: Strategies for Effective Oral Presentations.

C H A P T E R

7

Classifying Plant Leaves

■ ■ ■ ■ ■ **INTRODUCTION**

An important problem in many fields involves classifying items or individuals as belonging to one of two or more populations. A biologist classifies plants or animals according to their species or gender. A banker classifies credit card applications as being acceptable or not acceptable. A medical investigator classifies human hearts as being normal or abnormal. A criminal lawyer classifies potential jurors as being predisposed to convict, to acquit, or to be neutral toward his or her client. The branch of statistics dealing with these types of problems is known as classification.

To develop a classification strategy, the investigator takes samples from various populations. These samples, known as training samples, are used to train the investigator in classifying items from these populations. The investigator measures one or more predictor variables on each item in the training samples. As these predictor variables will be used to classify future items with regard to their population, the investigator wants to select predictor variables whose typical values tend to differ across the various populations. For example, height would be a better predictor variable than age for classifying college freshmen according to their gender.

With two predictor variables, each item in the training samples can be represented by a point in a plane—the observation space. With three predictor variables, the observation space becomes three dimensional. If the data clouds for the various training samples are separated in the observation space, then excellent classification rules can be developed. If there is considerable overlap in the data clouds, then classification rules based on these predictor variables will have difficulty in correctly classifying future items.

A geometric view is that we are partitioning the observation space into distinct regions corresponding to each population. For example, with two populations and two predictor variables, the classification rule splits the observation plane into two distinct parts. The predictor variables for a future item will be in the region corresponding to one of the populations. The classification rule will assert that the item came from that population. The statistical question is how to partition the observation space.

Computing Concepts and Procedures

SAS PROC DISCRIM, two-dimensional plot

Mathematical Concepts

Linear inequality, matrix operations

Statistical Concepts

Bivariate normal distribution, classification, density function, discriminant analysis, scatter plot, unbiased estimator

Materials Required

For each team, a ruler marked in millimeters

■ ■ ■ ■ ■ THE SETTING

You are a statistician employed in a college's statistical consulting unit. Biology major Iris Bloom asks you to develop a method for classifying leaves from a cherry tree species (Species A) and a pear tree species (Species B). She wants to use the method in the future to classify leaves with respect to these species. The available measurement device is a metric ruler.

■ ■ ■ ■ ■ THE BACKGROUND

We have two predictor variables, leaf blade width and length, denoted by X and Y, respectively. Your boss, Maria Thomas, suggests that you initially assume the width and length variables follow bivariate normal distributions for the two species with different mean vectors but equal covariance matrices. The bivariate normal density function depends on the mean parameters $\mu_x \in (-\infty, \infty)$ and $\mu_y \in (-\infty, \infty)$, the variance parameters $\sigma_x^2 > 0$ and $\sigma_y^2 > 0$, and the covariance parameter $\sigma_{xy} \in (-\sigma_x\sigma_y, \sigma_x\sigma_y)$. The density function is defined for all real x and y by

$$f(x,y) = \frac{1}{2\pi|\Sigma|^{1/2}} \exp\left[-\frac{1}{2}\begin{pmatrix} x-\mu_x \\ y-\mu_y \end{pmatrix}' \Sigma^{-1} \begin{pmatrix} x-\mu_x \\ y-\mu_y \end{pmatrix}\right] \quad (7.1)$$

with the covariance matrix

$$\Sigma = \begin{pmatrix} \sigma_x^2 & \sigma_{xy} \\ \sigma_{xy} & \sigma_y^2 \end{pmatrix} \quad (7.2)$$

▪ ▪ ▪ ▪ ▪ THE EXPERIMENT

Step 1: The Basic Measurement

The basic measurement will be to determine the width and length of a leaf blade to the nearest millimeter. Others will likely make the length and width measurements in the future, so we must make precise operational definitions of these variables. Define the base to be the point where the petiole (leafstalk) attaches to the blade (body) of the leaf. The length of the leaf blade will be the distance from the base to the most distant leaf tip. The width of the leaf blade will be the maximum distance across the leaf blade on a line perpendicular to the imaginary line connecting the base to the most distant leaf tip. Figure 7.1 illustrates these measurements for two types of leaves.

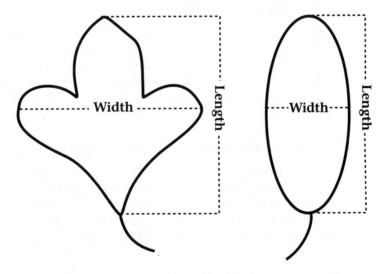

FIGURE 7.1 Leaf blade width and length for two types of leaves

TABLE 7.1 Leaf Blade Width and Length for Species A and B

SPECIES A			SPECIES B		
LEAF	WIDTH	LENGTH	LEAF	WIDTH	LENGTH
1			1		
2			2		
3			3		
4			4		
5			5		
6			6		
7			7		
8			8		
9			9		
10			10		
11			11		
12			12		
13			13		
14			14		
15			15		
16			16		

Step 2: Data Collection

Life-size images of the 16 leaves that Iris collected for each species appear in Figures 7.2 and 7.3. Number the leaf images from 1 to 16 for each species. Use a ruler to measure the blade width and length of each leaf to the nearest millimeter and record the data in Table 7.1. Be precise in your measurements. Measurement errors lead to less effective classification rules.

Step 3: Developing the Classification Rule

A reasonable approach for classification is to classify a future item as coming from the population with the larger density function value at the item's observed width and length—i.e., its (x,y) value. That is, we would classify a future leaf as coming from Species A if the leaf blade's width and length, (x,y), is a point such that

$$\frac{f_a(x,y)}{f_b(x,y)} \geq 1 \tag{7.3}$$

where f_a and f_b are the bivariate normal density functions for Species A and Species B, respectively. Otherwise, the future leaf would be classified as coming from Species B. Unfortunately, f_a and f_b depend on unknown parameters.

To distinguish between the mean parameters for the two species, let's add a second subscript to μ_x and μ_y to denote the species. Thus, μ_{xa} and μ_{ya} are the population means for Species A. The means for Species B are μ_{xb} and μ_{yb}. Recall that we are initially assuming that the covariance matrix Σ is the same for the two species.

We will use the data from the training samples to estimate these parameters and then define a classification rule in terms of estimates of f_a and f_b. Let

$$(X_{ia}, Y_{ia}), \ i = 1, \ldots, 16 \text{ and } (X_{ib}, Y_{ib}), \ i = 1, \ldots, 16 \qquad (7.4)$$

denote the training sample data from Species A and Species B, respectively. The sample means are

$$\overline{X}_a = \sum_{i=1}^{16} \frac{X_{ia}}{16}, \ \overline{Y}_a = \sum_{i=1}^{16} \frac{Y_{ia}}{16}, \ \overline{X}_b = \sum_{i=1}^{16} \frac{X_{ib}}{16}, \ \overline{Y}_b = \sum_{i=1}^{16} \frac{Y_{ib}}{16} \qquad (7.5)$$

Denote the sample variances for the two variables and the two species by

$$S_{xa}^2 = \sum_{i=1}^{16} \frac{\left(X_{ia} - \overline{X}_a\right)^2}{15} \qquad S_{ya}^2 = \sum_{i=1}^{16} \frac{\left(Y_{ia} - \overline{Y}_a\right)^2}{15}$$

$$S_{xb}^2 = \sum_{i=1}^{16} \frac{\left(X_{ib} - \overline{X}_b\right)^2}{15} \qquad S_{yb}^2 = \sum_{i=1}^{16} \frac{\left(Y_{ib} - \overline{Y}_b\right)^2}{15} \qquad (7.6)$$

The sample covariances are

$$S_{xya} = \sum_{i=1}^{16} \frac{\left(X_{ia} - \overline{X}_a\right)\left(Y_{ia} - \overline{Y}_a\right)}{15} \text{ and } S_{xyb} = \sum_{i=1}^{16} \frac{\left(X_{ib} - \overline{X}_b\right)\left(Y_{ib} - \overline{Y}_b\right)}{15} \qquad (7.7)$$

The mean parameters μ_{xa}, μ_{ya}, μ_{xb}, and μ_{yb} are estimated by the respective sample means $\overline{X}_a$, $\overline{Y}_a$, $\overline{X}_b$, and $\overline{Y}_b$. We will use a pooled sample estimate of the covariance matrix Σ analogous to that used to estimate the common variance in the two-sample pooled t-statistic. The estimate of Σ is

$$S = \begin{pmatrix} \dfrac{S_{xa}^2 + S_{xb}^2}{2} & \dfrac{S_{xya} + S_{xyb}}{2} \\ \dfrac{S_{xya} + S_{xyb}}{2} & \dfrac{S_{ya}^2 + S_{yb}^2}{2} \end{pmatrix} \qquad (7.8)$$

FIGURE 7.2 Images of cherry tree (Species A) leaves

FIGURE 7.2 Images of cherry tree (Species A) leaves *(continued)*

FIGURE 7.3 Images of pear tree (Species B) leaves

FIGURE 7.3 Images of pear tree (Species B) leaves *(continued)*

Replacing the parameters in f_a and f_b by their estimates leads to the rule of classifying a future leaf as coming from Species A if its (x,y) value satisfies

$$\frac{\hat{f}_a(x,y)}{\hat{f}_b(x,y)} = \frac{\dfrac{1}{2\pi|S|^{1/2}} \exp\left[-\dfrac{1}{2}\begin{pmatrix} x-\overline{X}_a \\ y-\overline{Y}_a \end{pmatrix}' S^{-1}\begin{pmatrix} x-\overline{X}_a \\ y-\overline{Y}_a \end{pmatrix}\right]}{\dfrac{1}{2\pi|S|^{1/2}} \exp\left[-\dfrac{1}{2}\begin{pmatrix} x-\overline{X}_b \\ y-\overline{Y}_b \end{pmatrix}' S^{-1}\begin{pmatrix} x-\overline{X}_b \\ y-\overline{Y}_b \end{pmatrix}\right]}$$

$$= \exp\left[-\frac{1}{2}\begin{pmatrix} x-\overline{X}_a \\ y-\overline{Y}_a \end{pmatrix}' S^{-1}\begin{pmatrix} x-\overline{X}_a \\ y-\overline{Y}_a \end{pmatrix} + \frac{1}{2}\begin{pmatrix} x-\overline{X}_b \\ y-\overline{Y}_b \end{pmatrix}' S^{-1}\begin{pmatrix} x-\overline{X}_b \\ y-\overline{Y}_b \end{pmatrix}\right] \geq 1 \tag{7.9}$$

and otherwise classifying the observation as coming from Species B.

Taking logarithms in (7.9) yields

$$-\frac{1}{2}\begin{pmatrix} x-\overline{X}_a \\ y-\overline{Y}_a \end{pmatrix}' S^{-1}\begin{pmatrix} x-\overline{X}_a \\ y-\overline{Y}_a \end{pmatrix} + \frac{1}{2}\begin{pmatrix} x-\overline{X}_b \\ y-\overline{Y}_b \end{pmatrix}' S^{-1}\begin{pmatrix} x-\overline{X}_b \\ y-\overline{Y}_b \end{pmatrix} \geq 0 \tag{7.10}$$

By distributing the multiplication, we have

$$-\frac{1}{2}\begin{pmatrix} x \\ y \end{pmatrix}' S^{-1}\begin{pmatrix} x \\ y \end{pmatrix} + \begin{pmatrix} \overline{X}_a \\ \overline{Y}_a \end{pmatrix}' S^{-1}\begin{pmatrix} x \\ y \end{pmatrix} - \frac{1}{2}\begin{pmatrix} \overline{X}_a \\ \overline{Y}_a \end{pmatrix}' S^{-1}\begin{pmatrix} \overline{X}_a \\ \overline{Y}_a \end{pmatrix}$$

$$+\frac{1}{2}\begin{pmatrix} x \\ y \end{pmatrix}' S^{-1}\begin{pmatrix} x \\ y \end{pmatrix} - \begin{pmatrix} \overline{X}_b \\ \overline{Y}_b \end{pmatrix}' S^{-1}\begin{pmatrix} x \\ y \end{pmatrix} + \frac{1}{2}\begin{pmatrix} \overline{X}_b \\ \overline{Y}_b \end{pmatrix}' S^{-1}\begin{pmatrix} \overline{X}_b \\ \overline{Y}_b \end{pmatrix} \geq 0 \tag{7.11}$$

Combining terms and rewriting gives the inequality

$$\begin{pmatrix} \overline{X}_a-\overline{X}_b \\ \overline{Y}_a-\overline{Y}_b \end{pmatrix}' S^{-1}\begin{pmatrix} x \\ y \end{pmatrix} \geq \frac{1}{2}\begin{pmatrix} \overline{X}_a+\overline{X}_b \\ \overline{Y}_a+\overline{Y}_b \end{pmatrix}' S^{-1}\begin{pmatrix} \overline{X}_a-\overline{X}_b \\ \overline{Y}_a-\overline{Y}_b \end{pmatrix} \tag{7.12}$$

Thus, our classification rule classifies a future leaf as coming from Species A if the item's (x,y) value satisfies Inequality (7.12). Otherwise, the item is classified as coming from Species B. The boundary between the two classification regions,

$$\begin{pmatrix} \overline{X}_a-\overline{X}_b \\ \overline{Y}_a-\overline{Y}_b \end{pmatrix}' S^{-1}\begin{pmatrix} x \\ y \end{pmatrix} = \frac{1}{2}\begin{pmatrix} \overline{X}_a+\overline{X}_b \\ \overline{Y}_a+\overline{Y}_b \end{pmatrix}' S^{-1}\begin{pmatrix} \overline{X}_a-\overline{X}_b \\ \overline{Y}_a-\overline{Y}_b \end{pmatrix} \tag{7.13}$$

defines a line in the plane.

TABLE 7.2 Summary Statistics for Example Training Samples

SPECIES A		SPECIES B	
$\overline{X}_a = 96.50$	$\overline{Y}_a = 119.10$	$\overline{X}_b = 86.75$	$\overline{Y}_b = 128.95$
$S_{xa}^2 = 14.8947$	$S_{ya}^2 = 59.9895$	$S_{xb}^2 = 22.6184$	$S_{yb}^2 = 44.2605$
$S_{xya} = 15.5789$		$S_{xyb} = 14.3553$	

Table 7.2 contains summary statistics for some example training samples from species different from those studied by Iris Bloom. The sample covariance matrix for the example data is

$$S = \begin{pmatrix} \dfrac{14.8947+22.6184}{2} & \dfrac{15.5789+14.3553}{2} \\ \dfrac{15.5789+14.3553}{2} & \dfrac{59.9895+44.2605}{2} \end{pmatrix} = \begin{pmatrix} 18.7566 & 14.9671 \\ 14.9671 & 52.1250 \end{pmatrix} \quad (7.14)$$

Substituting the sample means from Table 7.2 and S from Equation (7.14) into Inequality (7.12) yields the classification rule for the example data. A future leaf would be classified as coming from Species A if its (x,y) value satisfied

$$\begin{pmatrix} 96.50-86.75 \\ 119.10-128.95 \end{pmatrix}' \begin{pmatrix} 18.7566 & 14.9671 \\ 14.9671 & 52.1250 \end{pmatrix}^{-1} \begin{pmatrix} x \\ y \end{pmatrix}$$
$$\geq \frac{1}{2} \begin{pmatrix} 96.50+86.75 \\ 119.10+128.95 \end{pmatrix}' \begin{pmatrix} 18.7566 & 14.9671 \\ 14.9671 & 52.1250 \end{pmatrix}^{-1} \begin{pmatrix} 96.50-86.75 \\ 119.10+128.95 \end{pmatrix} \quad (7.15)$$

After matrix calculations, Inequality (7.15) can be written as

$$0.8699x - 0.4388y \geq 25.2903 \quad (7.16)$$

For the example data, a future leaf with a width of 100 mm and a length of 115 mm would be classified as coming from Species A because

$$0.8699(100) - 0.4388(115) = 36.5280 \geq 25.2903 \quad (7.17)$$

Figure 7.4 illustrates the partition of the plane on a scatter plot of the example training samples. Any future leaf with width and length on or below the line would be classified as coming from Species A. If the item's width and length fall above the line, the item would be classified as coming from Species B. Notice that the rule correctly classifies almost all training sample items.

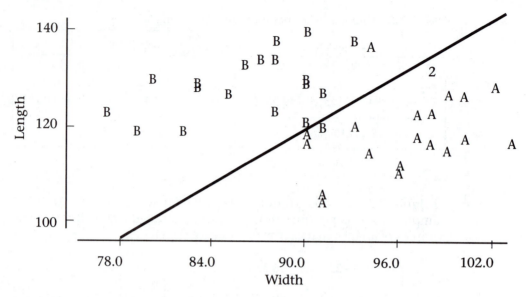

FIGURE 7.4 Partition of the plane for example data

The classification method we developed is known as linear discriminant analysis. This popular statistical tool is included in many statistical computing packages. The following SAS code performs linear discriminant analysis on the example training samples and classifies future items:

```
data one;
input width length type $;
cards;
103 117 A
 99 127 A
          An entry for each training sample item goes here.
 85 127 B
data two;
input width length;
cards;
100 115
          An entry for each "future" item to be classified goes here.
99 124
proc discrim data=one testdata=two method=normal pool
    =yes list testlist;
class type;
run;
```

■ ■ ■ ■ ■ PARTING GLANCES

We developed a classification rule using two populations and two predictor variables with equal sample sizes. With unequal sample sizes a different pooled estimate of Σ is used. It is also possible to extend these results to more than two populations and more than two predictor variables.

We have seen that linear discriminant analysis is a very logical classification rule under the assumption that the training samples come from bivariate (or multivariate) normal distributions with equal covariance matrices. Removing the assumption of equal covariance matrices or of normality leads to other classification rules.

If we assume normality but do not assume equal covariance matrices, we use separate estimates of the covariance matrix for each training sample. This leads to replacing the pooled estimate S in Equation (7.9) by the separate estimates. This change results in a curved classification boundary instead of the linear boundary given in Equation (7.13). This type of classification is known as quadratic discriminant analysis.

If we remove the normality assumption, the functional form of the ratio of density functions is unknown. A fundamentally different approach is required. Classification techniques not based on the normality assumption are known as nonparametric discriminant analysis.

In some classification problems, the investigator can express prior probabilities on the events that a future item will come from each of the populations. For example, Iris Bloom might believe that 90% of all future leaves will come from Species A and 10% will come from Species B. Thus, the prior probabilities that a leaf will be from Species A and B are .9 and .1, respectively. These prior probabilities can be used to compute posterior probabilities that a future item came from each of the populations after the predictor variables are measured for that item.

■ ■ ■ ■ ■ ASSIGNMENT: ANALYZING THE DATA

Include a copy of Table 7.1 with your responses.

1. Create a scatter plot of your training sample data, using the species letter as the plotting symbol. Do you have any outliers? If you have made gross measurement errors or data entry errors, correct the errors and redo the plot. If not, why are the outlier leaves so different from the rest?
2. Compute summary statistics for your two training samples and construct a table similar to Table 7.2.
3. Do you believe the assumption of bivariate normal distributions for the width and length is justified? Explain.
4. Do you believe the assumption of equal covariance matrices is justified? Explain.

5. Evaluate the classification rule in Inequality (7.12) using your summary statistics. Indicate the boundary on the scatter plot you created in Problem 1.
6. Use your classification rule to classify each observation in your training samples and a future leaf with 42-mm width and 68-mm length. Determine the number of correct and incorrect classifications for each training sample.
7. Use a discriminant analysis routine in a statistical computing package to classify each observation in your training samples and a future leaf with 42-mm width and 68-mm length. Do the results agree with your answers in Problem 6?
8. What further analysis, if any, do you suggest for these data? Why might this analysis be useful?

■ ■ ■ ■ ■ ASSIGNMENT: LOOKING AT THE MATHEMATICS

1. Let S_1 and S_2 denote separate estimates of the covariance matrices for Species A and Species B. Replace S in Equation (7.9) by S_1 and S_2 in the expressions for $\hat{f}_a(x,y)$ and $\hat{f}_b(x,y)$, respectively. Show that the resulting classification rule is not a linear inequality.

■ ■ ■ ■ ■ ASSIGNMENT: THE WRITTEN PRESENTATION

Turn in a copy of Table 7.1 separate from your written presentation.
In your role as a consulting center statistician, write a formal report summarizing your work to be read by biology major Iris Bloom, her advisor, and your boss. Your report should include the following:

1. Write a summary of the experiment that was performed and its purpose. Refer to the two species as cherry and pear.
2. A description of the two predictor variables, their operational definitions, and the measurement device
3. A statement of assumptions and an assessment of how reasonable they are
4. Summary statistics for each training sample
5. The classification rule and an example illustrating its use
6. A summary of the performance of the classification rule when classifying items from the training samples
7. Supply suggestions for future experiments or improvements to this experiment. These suggestions might include different predictor variables, measurement protocol, or analyses.
8. Professional-quality tables and figures, as necessary

For guidance, refer to Chapter 12: Strategies for Effective Written Reports.

■ ■ ■ ■ ■ ASSIGNMENT: THE ORAL PRESENTATION

In your role as a consulting center statistician, prepare a 5-minute oral presentation including professional-quality visual displays to present the results of this experiment to biology major Iris Bloom, her advisor, and your boss. Your presentation should address items 1–8 of The Written Presentation assignment section. For guidance, refer to Chapter 13: Strategies for Effective Oral Presentations.

C H A P T E R
8

Using a Response Surface to Optimize Product Performance

■ ■ ■ ■ ■ INTRODUCTION

Frequently, engineering factors can be adjusted over a range of possible settings to affect aspects of the performance of a product or process. For example, adjusting the latch position can affect the amount of force required to close a car door. A chemist can adjust the amounts of various chemicals to affect the yield of a chemical reaction. A baker can adjust the cooking time and temperature to affect the taste and texture of a cake. Statisticians and engineers are often faced with the task of searching for the combination of factor settings that optimizes product or process performance. In our examples, that would be searching for the latch position that minimizes the force required to close the door, the amounts of the chemicals that give the largest chemical reaction yield, and the cooking times and temperatures that produce the best cake.

The problem is most easily envisioned for the case of adjusting two factors, A and B, whose possible values are all points in an interval. In this case, the expected product performance measure can be thought of as being on a surface above the AB-plane. Statisticians refer to this surface of expected responses as a response surface. Response surface techniques are an important branch of multiple regression and experimental design. Figure 8.1 illustrates a response surface that achieves a maximum with $A = 3$ and $B = 2$.

Unfortunately, in real applications, no one knows what the true response surface is. The statistician and the engineer work together to design and perform a series of experiments. They fit a model to the data to estimate the response surface. This estimated surface is called the sample

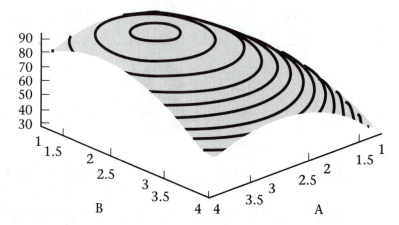

FIGURE 8.1 An example response surface

response surface. Finding and confirming optimal factor settings often require several experiments. This is especially true if the factor settings in the initial experiment are far from optimal.

Response surface techniques can have a major impact in industry—perhaps the difference between a company's success and failure. The maximum expected response in Figure 8.1 is three times the minimum. A company operating its process near the optimal settings may be very profitable, whereas a competitor operating far from the optimal settings may be driven out of business. Which of these companies do you want to work for? Statisticians can have a major impact on a company's performance.

Computing Concepts and Procedures

SAS PROC GLM, SAS PROC REG, three-dimensional plot

Mathematical Concepts

Maximization, quadratic surface, system of linear equations

Statistical Concepts

Experimental design, factor selection, planned experiment, multiple regression, randomization, response surface.

Materials Required

For each team, a balsa wood airplane with movable wings, a ruler, four paper clips, and a 50-foot measuring tape

▪ ▪ ▪ ▪ ▪ THE SETTING

You are a statistician employed by Flying B Aerospace, a leading producer of balsa wood airplanes. Aeronautical engineer Bruce Wright has come to you for assistance. He wants to find the wing position and nose weighting that will achieve the maximum flight distance for the experimental model X234J airplane. His operational definition of flight distance is the distance from the launch site to the closest point on the landed airplane. His product development budget will allow at most 36 test flights.

▪ ▪ ▪ ▪ ▪ THE BACKGROUND

The body of the model X234J balsa wood airplane has a slot through which the wings are inserted. The slot is longer than the width of the wings. This allows for adjustment of the wing position relative to the front of the airplane. A metal nose weight is attached to the front of the plane body. The airplane body and wing assembly are illustrated in Figure 8.2.

Bruce Wright's experience with similar models indicates that both the wing position and the amount of nose weight affect flight distance and that the two factors interact. Moving the wings forward causes the nose to lift in flight; increasing the nose weight tends to pull the nose down. A slight elevation of the nose can result in long stable flights. A large elevation of the nose causes an airplane to soar and then stall. Although this usually produces a short flight, it occasionally produces an unusually long flight if the airplane encounters favorable wind currents. Considerable variation in flight distances can be expected for any combination of wing position and nose weight.

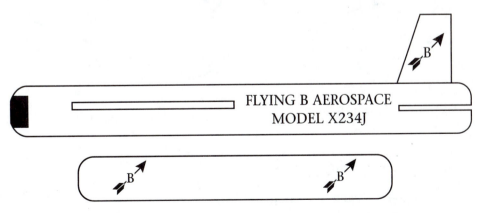

FIGURE 8.2 Body and wing assembly of model X234J balsa wood airplane

The popular quadratic response model expresses the flight distance, *Y*, as a second-order polynomial in the factors plus a random error term,

$$Y = \beta_0 + \beta_1 A + \beta_2 B + \beta_3 AB + \beta_4 A^2 + \beta_5 B^2 + \varepsilon \tag{8.1}$$

Fitting the response surface is equivalent to estimating the six unknown β parameters. The model is quite general and will fit a wide variety of shapes. In some applications, transformation of the response (dependent) variable or the factors will improve the fit. The quadratic response model may break down if the (*A*, *B*) region of interest is large.

In designing your experiment, plan for at least two stages of data collection. The first stage is used to get a basic understanding of the shape of the surface. The design of the additional stage or stages depends on the first stage data. In industrial experiments, there are typically several stages of experimentation.

To understand the importance of collecting data in stages, consider the sample response surface in Figure 8.3. The contours suggest that the optimal factor settings are well outside the data collection region. We might suspect that the optimal factor settings have $A < 1$ and $B > 4$. We need to collect more data. If we allocated all of our observations to the first stage, collecting more data would not be an option. If our sample response surface is like that in Figure 8.1, we would like to take additional observations in the neighborhood of $A = 3$ and $B = 2$ to fine-tune our estimate of the optimal settings.

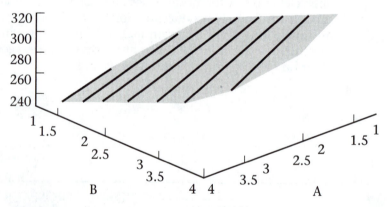

FIGURE 8.3 Sample response surface far from maximum

■ ■ ■ ■ ■ THE EXPERIMENT

Step 1: The Basic Measurement

The basic measurement is the shortest distance (in inches) from the launch site to the closest point on the landed airplane for each flight.

Step 2: Experimental Design for the First Stage

The experimental design for the first stage involves specifying the number of flights and the combinations of wing position and nose weight that will be used on the flights. A total of 36 flights are possible, so it would be reasonable to use about half of these, or 18, in the first stage of testing. This will leave the rest for further study after you learn more about the surface.

Let factor A dictate the distance (in inches) from the front of the slot to the wing's front edge. It is convenient to mark a scale on the airplane to correctly position the wings for each flight. Figure 8.4 illustrates such a scale. Let factor B dictate the number of paper clips that are placed on the nose.

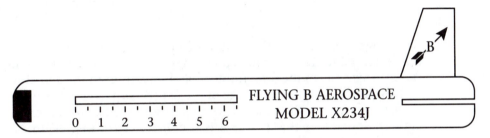

FIGURE 8.4 Scale for determining wing position (factor A)

In choosing the wing position and nose weight combinations, you must ensure that you will be able to estimate all unknown parameters and that you will be able to assess the fit of the quadratic response surface model. Industrial experiments are often time consuming and expensive. **You do not want to learn that the experimental design is inadequate after the data have been collected.** Two commonly used designs that allow for parameter estimation and an assessment of fit are the replicated factorial design and the central composite design with replication. Let's look at examples of these types of designs.

Let $A_1 < A_2 < A_3 < A_4 < A_5$ denote evenly spaced, ordered values of factor A (the wing position). For example, A_i might represent that the front of the wing is positioned $i - 1$ units from the front of the wing slot. Let $B_1 < B_2 < B_3 < B_4 < B_5$ denote evenly spaced, ordered values of factor B, (the nose weight). For example, B_j might represent the addition of $j - 1$ paper clips to the nose of the airplane.

An example of a replicated factorial design would be to make two flights with each of the following nine treatment combinations: (A_1, B_1), (A_1, B_3), (A_1, B_5), (A_3, B_1), (A_3, B_3), (A_3, B_5), (A_5, B_1), (A_5, B_3), and (A_5, B_5). An example of a central composite design with replication would be to make two flights with each of the following nine treatment combinations: (A_1, B_3), (A_2, B_2), (A_2, B_4), (A_3, B_1), (A_3, B_3), (A_3, B_5), (A_4, B_2), (A_4, B_4), and (A_5, B_3). The squares in Figure 8.5 illustrate the design points for these

Factorial   Central Composite

FIGURE 8.5 Design points for factorial and central composite designs

two designs. In practice, the central composite design is often run with the center point (A_3, B_3) being replicated several times and no other replication. This reduces the total number of observations in the experiment.

Now let's design the first stage of the experiment. Use a pen and ruler to mark a measurement scale for wing position (see Figure 8.4 for an example). Remember that factor A is the distance (in inches) from the front of the slot to the front edge of the wing. The minimum setting of factor A is 0, which implies that the front edge of the wing touches the front of the slot. The maximum setting of factor A occurs when the back edge of the wing touches the back of the slot. This maximum value depends on the length of the slot and the width of the wing. Now, choose five evenly spaced locations on this scale to serve as levels of factor A. These values should be ordered such that $0 \leq A_1 < A_2 < A_3 < A_4 < A_5 \leq$ maximum possible value of A. Mark these locations on the body of the airplane and complete the descriptions of the levels of factor A in Table 8.1. The levels of factor B will indicate different numbers of paper clips being added to the nose.

TABLE 8.1 Levels of Factors A and B

FACTOR A	FRONT OF SLOT TO FRONT OF WING (IN INCHES)	FACTOR B	NUMBER OF PAPER CLIPS
A_1		B_1	0
A_2		B_2	1
A_3		B_3	2
A_4		B_4	3
A_5		B_5	4

Now choose your design. You can use the replicated factorial design, the central composite design with replication, or the central composite design with replication only at the center point. Unless you have studied experimental design, your choice among these designs may be quite arbitrary. You are encouraged to use no more than 18 flights in the first stage. Write the treatment combinations you will use in the top rows of Table 8.2.

There are a number of factors in addition to wing position and nose weight that affect flight distance. These factors include the height, velocity, and angle of the launch, and weather conditions (such as wind currents, temperature, and humidity). You should attempt to keep the launch conditions as consistent as possible throughout the flights. This is best done by having a single person launch all flights and having him or her tuck the launching elbow against his or her side as the airplane is launched with a flip of the wrist. Despite your best efforts at consistency, some variation will remain in these launch variables. You will also have no control over the weather variables unless you do the experiment in a controlled environment, such as a gym. The planes are designed to be flown outside, so a controlled environment is not entirely satisfactory.

To ensure that all treatment combinations have equal access to the various launch and weather conditions occurring during the first stage, we will randomize the order of your first stage flights. Locate a source of random numbers. Most statistical computing packages will generate random numbers. If such a package is not available, use a random number table or the random function on a calculator. You will use the random numbers to randomize the order of the first stage flights. For example, if you have 18 flights in the first stage, you would generate random integers between 1 and 18. The random number stream "3, 13, 14, 1, 9, 6, 7, . . ." would indicate that the treatment combination listed in row 3 of Table 8.2 would have flight order = 1 and be flown first, the treatment combination listed in row 13 would have flight order = 2 and be flown second, and so on. Use your random numbers to establish the order of the first stage flights and write the flight order values in the fourth column of Table 8.2. Ignore repeat occurrences of a number in the random number stream.

Step 3: Data Collection

Take your airplane, 4 paper clips, a 50-foot measuring tape, and a copy of Table 8.2 to the launch location. Before launching the airplane, **make sure the airplane is configured according to the experimental design for that flight.** That is, before launching the first flight (flight order = 1), make sure that the front of the wings is at the mark indicating wing position for the level of factor *A* to be used on the flight with flight order 1. In a similar fashion, make sure that the number of paper clips on the

TABLE 8.2 Experimental Design and Flight Distances

FLIGHT NUMBER	FACTOR A LEVEL	FACTOR B LEVEL	FLIGHT ORDER	FLIGHT DISTANCE
1				
2				
3				
4				
5				
6				
7				
8				
9				
10				
11				
12				
13				
14				
15				
16				
17				
18				
19				
20				
21				
22				
23				
24				
25				
26				
27				
28				
29				
30				
31				
32				
33				
34				
35				
36				

nose matches that specified for the flight with flight order 1. Now, make sure the flight path is clear and launch the airplane. **Remember that the launcher should keep the launching elbow tucked to his or her side.** The data collectors will measure the flight distance using the operational definition—the distance from the launch site to the closest point on the landed airplane. Record the flight distance in the fifth column in the row of Table 8.2 corresponding to the flight. Repeat this process for all flights in the first stage using the order specified in the fourth column of Table 8.2.

Step 4: Fitting the Model

Now we are ready to fit the quadratic response surface model in Equation (8.1). Use a statistical computing package to fit the response surface model to your data. This is often done using the multiple-regression portion of a statistical computing package. You will input the following information for each flight:

> The number of inches from the front of the slot to the front edge of the wings, A
> The number of paper clips placed on the nose, B
> The flight distance (in inches), Y

Report the results of your fitted model in Table 8.3.

TABLE 8.3 Response Surface Parameter Estimates and Analysis of Variance Table

SOURCE	DEGREES OF FREEDOM	SUM OF SQUARES	MEAN SQUARE	F-VALUE	p-VALUE	r-SQUARE
Model						
Error						
Corrected total						

VARIABLE	COEFFICIENT ESTIMATE	STANDARD ERROR	t-STATISTIC	p-VALUE
Intercept				
A				
B				
AB				
A^2				
B^2				

The following SAS code fits the quadratic response surface for this problem:

```
data one;
input a b y;
a2=a*a;
b2=b*b;
ab=a*b;
cards;
0.00 0 123.25
```
                    An entry for each flight goes here.
```
6.00 4 53.00
proc print;
proc reg;
model y = a b ab a2 b2;
run;
```

Step 5: Assessing the Fit

The coefficient estimates in Table 8.3 reflect the sample response surface

$$\hat{Y} = \hat{\beta}_0 + \hat{\beta}_1 A + \hat{\beta}_2 B + \hat{\beta}_3 AB + \hat{\beta}_4 A^2 + \hat{\beta}_5 B^2 \tag{8.2}$$

Among all quadratic response surfaces, this sample response surface is best in that it maximizes the sum of squares model and minimizes the sum of squares error (SS_{error}) over the class of all quadratic response surfaces. We do not know that the true response surface is quadratic, so it is wise to assess the fit of the quadratic response model before we make inferences based on this sample response surface.

Because we have made more than one flight for at least one treatment combination, it will be possible to do a formal test of fit of our model. We will test the null hypothesis that the response surface model is quadratic (Equation 8.1) versus the alternative hypothesis that the model is not quadratic. To do this test, we partition the degrees of freedom for error and sum of squares error in Table 8.2 into two parts

$$df_{error} = df_{lack\ of\ fit} + df_{pure\ error}$$
$$SS_{error} = SS_{lack\ of\ fit} + SS_{pure\ error} \tag{8.3}$$

Denote the number of flights in the first stage by n and the number of different treatment combinations used by k. For the factorial design and the central composite design depicted in Figure 8.5, $k = 9$. Determine the values of n and k for your first-stage experiment. Complete the entries for the degrees of freedom and the sum of squares error in Table 8.4.

Consider the set of flight distances for all flights using the treatment combination (A_i, B_j). Let SS_{ij} denote the numerator of the sample variance

TABLE 8.4 Test of Fit Analysis

SOURCE	D.F.	SUM OF SQUARES	MEAN SQUARE	F-VALUE	p-VALUE
Error	$n - 6 =$				
Lack of fit	$k - 6 =$				
Pure error	$n - k =$				

of these distances. The $SS_{\text{pure error}}$ is the sum of the SS_{ij} terms over all treatment combinations. The $SS_{\text{pure error}}$ does not depend on the model unless we transform Y. The $SS_{\text{lack of fit}} = SS_{\text{error}} - SS_{\text{pure error}}$ does depend on the model. The worse the model fits the data, the larger SS_{error} and $SS_{\text{lack of fit}}$ will be.

One way to compute $SS_{\text{pure error}}$ involves fitting a two-way analysis of variance model with interaction. The $SS_{\text{pure error}}$ equals the sum of squares error from this analysis of variance. This analysis of variance model puts no restrictions on the means, hence the $SS_{\text{lack of fit}}$ from the quadratic response surface fit is explained by the two-way analysis of variance model with interaction.

The following SAS code performs this analysis of variance:

```
data one;
input a b y;
cards;
0.00 1 123.25
                An entry for each flight goes here.
6.00 4 53.00
proc print;
proc glm;
class a b;
model y = a b a*b;
run;
```

Compute the $SS_{\text{pure error}}$ and $SS_{\text{lack of fit}}$ and write the values in Table 8.4. The mean squares for lack of fit and pure error are found by dividing each sum of squares by its degrees of freedom. Compute the mean squares and write them in Table 8.4. The test statistic for lack of fit is

$$F = \frac{MS_{\text{lack of fit}}}{MS_{\text{pure error}}} \tag{8.4}$$

Under the assumption that the model (8.1) is correct and that the random error terms are independent, identically distributed normal random variables, the test of fit statistic has an F distribution with $k - 6$ and $n - k$ degrees of freedom. Large values of F suggest that the quadratic response model is incorrect. Compute the test of fit statistic and write the result in Table 8.4. Compute the *p*-value

$$P(F_{k-6,n-k} \geq \text{observed } F)$$

for your test and write it in Table 8.4. This *p*-value is one minus the $F_{k-6,n-k}$ distribution function evaluated at the observed value of F. What does this *p*-value suggest about your model?

If you are satisfied with your model, move on to step 6. Otherwise, just say "Drat," and then repeat steps 4 and 5 with a different model. Looking at the residuals from the fit of the quadratic response surface should help in developing a new model. If your new model transforms the Y variable, you will need to recompute the $SS_{\text{pure error}}$. If your new model uses a different number of regression coefficients, you will need to replace the 6 in Table 8.4 and in the numerator degrees of freedom for the F statistic by the number of regression coefficients.

The residuals and other regression diagnostics can also be helpful in identifying unusual data points, known as outliers. Outliers can be caused by data entry errors, measurement errors, incorrect models, and unusual experimental conditions. These data points should be checked with care.

Step 6: Exploring the Sample Response Surface

Now that you are at least reasonably happy with the fitted model, you need to learn about the shape of the sample response surface and determine the factor settings where the sample response surface has its maximum height. To begin, produce a three-dimensional plot of the sample response surface over the region in the *AB*-plane from which you collected data. We will refer to this region as the experimental region.

Figures 8.1 and 8.3 are examples of three-dimensional plots made by using Maple® V. The Maple V command

```
plot3d(30 + 4*A + 6*B - A*B + 2*A^2 - 3*B^2, A = 1..5,
    B = 1..5);
```

produces a three-dimensional plot of the sample response surface

$$\hat{Y} = 30 + 4A + 6B - AB + 2A^2 - 3B^2$$

over the square region with $1 \leq A \leq 5$ and $1 \leq B \leq 5$. The Contour option of SAS PROC PLOT can be used if a three-dimensional plotting routine is not available.

Look at your three-dimensional plot. Do you see evidence of a local maximum in the sample response surface within your experimental region? If you do, then the optimal factor settings are probably close to the settings where the sample response surface reaches its peak.

If your three-dimensional plot does *not* show evidence of a local maximum in the sample response surface within your experimental region, just say "Double drat." In this case, the optimal factor settings may be far away from the factor settings used in the first stage.

To determine an estimate of the factor settings that optimize the flight distance, you need to find the values of A and B that maximize the sample response surface. This is done by setting the partial derivatives of the sample response surface with respect to A and B equal to 0 and solving for A and B. For the quadratic sample response surface given in Equation (8.2), these partial derivative equations are

$$\frac{\partial \hat{Y}}{\partial A} = \hat{\beta}_1 + \hat{\beta}_3 B + 2\hat{\beta}_4 A = 0$$

(8.5)

$$\frac{\partial \hat{Y}}{\partial B} = \hat{\beta}_2 + \hat{\beta}_3 A + 2\hat{\beta}_5 B = 0$$

Of course, you will need to verify that the solution does in fact yield a maximum. This is accomplished by showing that

$$\frac{\partial^2 \hat{Y}}{\partial A^2} \frac{\partial^2 \hat{Y}}{\partial B^2} - \left(\frac{\partial^2 \hat{Y}}{\partial A \partial B} \right)^2 > 0 \text{ and that either } \frac{\partial^2 \hat{Y}}{\partial A^2} < 0 \text{ or } \frac{\partial^2 \hat{Y}}{\partial B^2} < 0 \quad (8.6)$$

Find the values of A and B that maximize your sample response surface.

At times, the values of A and B that maximize the sample response surface are impossible values. For example, it is not possible to put a negative number of paper clips on the front of the airplane. If the values of A and B that maximize your sample response surface are not possible, they may still indicate the direction that you must move from the experimental region to find the optimal factor settings.

Step 7: Experimental Design for the Next Stage

You are now ready to plan the next stage of your experiment. As in the previous stage, the experimental design involves specifying the number of flights and the wing position and nose weight combinations that will be used in the flights.

The number of flights in this stage and the factor settings will depend on the results of the previous stage(s). If the sample response surface from the previous stage achieves a maximum in or near the experimental region, you will probably want to use all of your remaining flights in this stage. Your flights on this stage will use factor settings similar to those in the previous stage(s) to fine-tune your estimates of the optimal factor settings.

If the maximum of the sample response surface occurs far outside the experimental region, then the estimates of the optimal factor settings are highly suspect. In this case, you may want to only use approximately half of your remaining flights in this stage. You will want the flights in this stage to use factor settings different from those used in the previous stage(s). It is recommended that the experimental region for this stage be located between the previous experimental region and the location of the maximum of the sample response surface. The current estimate of the optimal factor settings is highly suspect, so gradual movement of the experimental region toward the estimated optimal settings may prove more effective than centering the new experimental region about the estimated optimal settings.

Follow the instructions in step 2 to determine the levels of factors A and B that will be used in your flights for this stage and to randomize the flight order. The definition of factor B suggests that it must be a nonnegative integer. You could approximate $B = 1.5$ by using one regular-size paper clip and another paper clip half the size of the regular paper clip. Collect the data for this stage by following the instructions in step 3. Repeat the analyses in steps 4–6 using the data from this and all previous stages.

If you have not yet used all 36 possible flights, you have the option of performing another stage of the experiment. To do this, return to the beginning of this step. Otherwise, your experimentation is complete and it is time to report the results to Bruce Wright.

■ ■ ■ ■ ■ **PARTING GLANCES**

Experimental design texts frequently use a shorthand system for identifying factor levels. If a factor has three levels, the lowest level is denoted by –, the middle level by 0, and the highest level by +. An ordered pair can be used to denote the levels of factors A and B in a treatment combination. Using this system, the nine treatment combinations in the factorial design shown in Figure 8.5 could be listed as (–,–), (–,0), (–,+), (0,–), (0,0), (0,+), (+,–), (+,0), and (+,+).

It is possible that the maximum of the sample response surface appears at levels of factors A and B that are difficult to achieve. For example, suppose that the sample response surface achieves its maximum when $A = -1$ and $B = -1$. This suggests that the wing should be further forward than the existing slot allows and that weight should be taken away from the nose rather than adding paper clips. These actions require more major modifications than were anticipated for this experiment. A knife or a saw should be used to extend the wing slot toward the front of the airplane and an appropriate tool should be used to remove some weight from the nose. Depending on the tools available to you and on the original design of the airplane, these actions may not be feasible. It is common in response surface experiments to have some combination of factor levels that are unachievable. At times, the optimal levels are on the boundary of the achievable factor levels.

The test of fit of the response surface model differs from most hypothesis tests. Generally, we wish to establish the alternative hypothesis by showing that the null hypothesis is false beyond a reasonable doubt. In the test of fit, the null hypothesis is that the response surface model is correct. We can never prove a null hypothesis, so we cannot prove that our model is correct. We either reject the null hypothesis and conclude our model is wrong, or we fail to reject the null hypothesis and conclude that we cannot rule out our model. In any experiment there are several models that cannot be ruled out. Even if our model is not exactly right, it may still prove very useful in helping us improve performance.

In an industrial experiment, we might add a blocking variable to the response surface model to allow for possibly different uncontrollable conditions in the various stages. To do this in SAS, we would add a new variable named *stage* to this data set and enter this variable into the model as a class variable. This would reduce the degrees of freedom for error by the number of stages minus one.

A different approach to maximization involves making the observations sequentially using a method (such as steepest ascent) to select the factor levels for the next observation.

At times, we wish to minimize a response variable, *Y*, rather than maximize it. In this case, we only need to change the signs of all observed responses.

■ ■ ■ ■ ■ ASSIGNMENT: ANALYZING THE DATA

Include copies of Tables 8.1, 8.2, 8.3, and 8.4 with your responses.

1. What experimental design did you use for the first stage?
2. What is the equation for the sample quadratic response surface that you fit to the data in the first stage?
3. What is the value of the *F* statistic for the test of fit of the quadratic response surface model for the first-stage data? Based on this *F* statistic, did you conclude that the quadratic response surface model was reasonable for modeling flight distance? If not, explain other model(s) that you tried and your thoughts regarding the adequacy of these models.
4. Based on the model that you fit to the data in the first stage, what wing position and number of paper clips appear to be optimal?
5. How did the outcome of the first stage of the experiment affect how you collected the data for the second stage?

6. Give the equation for the sample response surface you fit to all of your data. Based on this model, what wing position and number of paper clips appear to be optimal?

7. What, if anything, would you do differently if you were to redo the experiment?

■ ■ ■ ■ ■ ASSIGNMENT: LOOKING AT THE MATHEMATICS

1. Write the equations in Equation (8.5) in matrix notation $Hx = h$ where $x' = [A, B]$. What are necessary and sufficient conditions on the coefficients of the sample response surface to yield a solution to these equations?

2. What are necessary and sufficient conditions on the coefficients of the sample response surface to yield a unique solution to Equation (8.5)?

3. What are necessary and sufficient conditions on the coefficients of the sample response surface so that the unique solution in Problem 2 maximizes the surface?

■ ■ ■ ■ ■ ASSIGNMENT: THE WRITTEN PRESENTATION

Turn in copies of Tables 8.1 and 8.2 separate from your written presentation.

In your role as a statistician for Flying B Aerospace, write a formal report summarizing your work to be read by aeronautical engineer Bruce Wright, his boss, and your boss. Your report should include the following:

1. A summary of the experiment that was performed and its purpose
2. A description of the factors, including the operational definition of the factor levels
3. The operational definition of flight distance and a description of the measurement protocol
4. The rationale for doing the experiment in stages
5. A description of the first stage, including the experimental design, the model, the sample response surface, the test of fit, and the estimate of the factor levels that maximize the sample response surface
6. A description of the second and subsequent stages of the experiment, including the experimental design and rationale for its use, the sample response surface, and the estimate of the factor levels that maximize the sample response surface
7. Interpretations of the statistical results and recommended actions to be considered based on these results

8. Suggestions for future experiments or improvements to this experiment
9. Professional-quality tables and figures, as necessary

For guidance, refer to Chapter 12: Strategies for Effective Written Reports.

■ ■ ■ ■ ■ **ASSIGNMENT: THE ORAL PRESENTATION**

In your role as a statistician for Flying B Aerospace, prepare a 5-minute oral presentation including professional-quality visual displays to present the results of this experiment to aeronautical engineer Bruce Wright, his boss, and your boss. Your presentation should address items 1–9 of The Written Presentation assignment section. For guidance, refer to Chapter 13: Strategies for Effective Oral Presentations.

C H A P T E R
9

Modeling Breaking Strength with Dichotomous Data

■ ■ ■ ■ ■ **INTRODUCTION**

An important quality characteristic in many manufacturing operations is product strength. Strength is important in items produced from a variety of materials, including fibers, paper, cardboard, concrete, and metals. An item's breaking strength is the minimum amount of stress required to make the item fail. As with most quality characteristics, breaking strength varies from item to item. It is often important to investigate the breaking strength distribution for manufactured items.

Engineers frequently conduct breaking strength experiments on sample items from a manufacturing process. In some experiments, each test item is subjected to continuously increasing stress until it fails. The amount of stress on the object when it fails is recorded as the item's breaking strength. This type of experiment often requires sophisticated equipment to continuously increase the stress and to precisely measure the breaking strength. Stress is continuously increased until the item fails, so this type of breaking strength data is continuous.

A different type of experiment involves subjecting each test item to a specific stress amount and observing whether the item fails. The data are dichotomous—items fail or they do not fail. If an item fails, we know the item's breaking strength is less than or equal to the stress applied. If an item does not fail, we know the item's breaking strength is greater than the stress applied. Although each item is subjected to a single stress amount, different stress amounts are used for different items.

A dichotomous data experiment typically requires less sophisticated measurement devices than a continuous data experiment. Consequently, the dichotomous data experiment is often simpler and cheaper to perform. However, the continuous data experiment provides more information about the breaking strength distribution because it produces a precise breaking strength measurement. The dichotomous data experiment produces only an upper or lower bound for each item's breaking strength. Therefore, more observations are needed with a dichotomous data experiment to obtain the same information obtained with a continuous data experiment.

Computing Concepts and Procedures

Programming Newton's method, SAS PROC LOGISTIC

Mathematical Concepts

Maximization, Newton's method, partial derivative, system of nonlinear equations

Statistical Concepts

Bernoulli trial, dichotomous data, goodness of fit, likelihood function, likelihood ratio test, logistic distribution, logistic regression, maximum likelihood estimation, scatter plot

Materials Required

For each team, 99 two-ply facial tissues with a minimum dimension of at least 8 inches, two 7-inch embroidery hoops, three full 12-ounce soft drink cans, a ruler marked in centimeters, and a 1-ounce egg-shaped fishing weight

■ ■ ■ ■ ■ THE SETTING

You are a statistician employed by the Cry Your Eyes Out Tissue Company. Process engineer Sheila McDuff wishes to investigate the breaking strength distribution of tissues produced on an experimental production line. The plant manager has told her that no funds are available to purchase the expensive equipment required to make precise measurements of breaking strength. Sheila has asked you to help her perform a dichotomous data experiment for breaking strength, to analyze the data, and to make a formal report. Based on some past studies, Sheila believes that the breaking strengths of the tissues may be well approximated by a logistic distribution. Your boss says that several distributions often provide reasonable approximations and recommends trying Sheila's suggestion of the logistic distribution first.

▪ ▪ ▪ ▪ ▪ THE BACKGROUND

The logistic density function, which depends on the parameters $\alpha \in (-\infty,\infty)$ and $\beta > 0$, is

$$f_{\alpha\beta}(x) = \frac{\beta_{\exp}(\alpha + \beta x)}{\left[1 + \exp(\alpha + \beta x)\right]^2}, \qquad -\infty < x < \infty \tag{9.1}$$

This density function is similar in shape to the normal density function except the logistic density has somewhat heavier tails. The distribution function is

$$F_{\alpha\beta}(x) = \frac{\exp(\alpha + \beta x)}{1 + \exp(\alpha + \beta x)} \tag{9.2}$$

The logistic distribution's mean and variance are $-\alpha/\beta$ and $\pi^2/(3\beta^2)$, respectively. The distribution function is an s-shaped increasing function. The parameter α determines the value of $F_{\alpha\beta}(0)$ and the parameter β determines how quickly $F_{\alpha\beta}$ rises toward 1. Figure 9.1 illustrates the logistic density and distribution functions for $\alpha = -9$ and $\beta = 1$.

Let T represent a randomly selected tissue's breaking strength. The probability that a randomly selected tissue will fail when subjected to a stress x is equal to $P(T \le x)$; the breaking strength random variable's

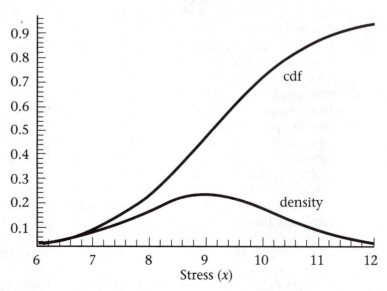

FIGURE 9.1   Logistic density and distribution functions, $\alpha = -9$ and $\beta = 1$

distribution function evaluated at x. If the breaking strengths follow a logistic distribution, then the probability that a randomly selected tissue will fail when subjected to a fixed stress x equals $F_{\alpha\beta}(x)$ for some unknown values of α and β. The probability that the tissue will not fail is given by

$$1 - F_{\alpha\beta}(x) = \frac{1}{1 + \exp(\alpha + \beta_x)} \qquad (9.3)$$

Expressing the probability of failure for varying stress amounts as the distribution function of the logistic distribution is known as logistic regression.

We need to test several tissues at a fixed stress amount and repeat the experiment for several stress amounts with other tissues to evaluate the logistic distribution breaking strength assumption and to estimate the unknown parameters α and β. We will test 16 tissues at each of 4 stress amounts for a total of 64 tissues. Stress will be measured by the distance (in centimeters) from which a 1-ounce fishing weight is dropped on the tissue.

■ ■ ■ ■ ■ THE EXPERIMENT

Step 1: The Basic Measurement

The basic measurement is determining whether a 1-ounce, egg-shaped fishing weight dropped from a predetermined distance falls through a two-ply facial tissue that is firmly clamped in a 7-inch embroidery hoop. The embroidery hoop is elevated on three 12-ounce soft drink cans. Begin the measurement process by clamping the tissue in the embroidery hoop. To do this, separate the outer and inner hoops (see Figure 9.2); it may be necessary to turn the clamp screw counterclockwise. Next, cover the inner hoop with the tissue, then place the inner hoop inside the outer hoop. Gently pull each tissue corner to make the tissue taut. Turn the clamp screw clockwise to tighten the outer hoop, making sure the tissue does not wrinkle or become loose. If the tissue becomes loose, loosen the clamp screw and adjust the tissue. If the tissue tears during clamping, discard it and use another. After you successfully clamp the tissue, place the hoop on the three soft drink cans (see Figure 9.3).

The next step of the measurement protocol is placing a ruler perpendicular to the surface of the tissue and then holding the weight next to the ruler such that the lowest part of the weight is the predetermined distance from the tissue. Figure 9.4 illustrates a weight 9 cm above the tissue. After releasing the weight, the measurement is either "yes" the weight fell through the tissue or "no" it did not. **Replace the tissue after each trial whether or not the tissue tears.**

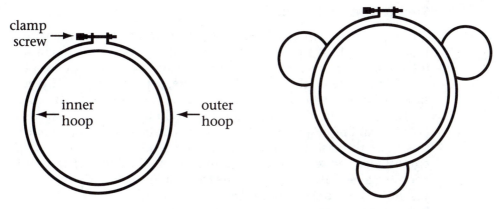

FIGURE 9.2 Embroidery hoop

FIGURE 9.3 Embroidery hoop on cans

To be sure your team understands the measurement protocol, drop the weight from 9 cm on each of three practice tissues. These three tissues are for practice only and will not be used in the data analysis. If you are comfortable with the measurement protocol after the practice measurements, you are ready for step 2. Otherwise, clarify any uncertainty with your team members and your instructor. Data collection moves faster if one team member places tissues on the hoops while another member performs the test. To minimize measurement variation, only one person should put the tissues on the hoops and only one person should drop the weights. Take your time collecting the measurements.

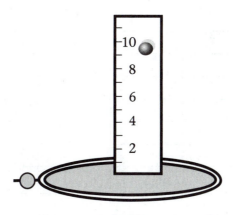

FIGURE 9.4 A weight 9 cm above tissue

Step 2: Data Collection

The experiment will consist of making the basic measurements on 16 tissues for each of 4 predetermined heights—7, 8, 9, and 10 cm—for a total of 64 measurements. Record the yes and no results for each tissue in Table 9.1. After all trials are complete, separately total the yes responses for each column and write them in the bottom of Table 9.1.

The maximum likelihood estimation technique requires that we have both yes and no responses for at least two distances. If you do not have this, make the basic measurement on 16 additional tissues using a different distance and record the results in an unused column of Table 9.1. In choosing a new distance, select one that you think is likely to yield both yes and no responses. Repeat this process as necessary until you have both yes and no responses for at least two distances.

TABLE 9.1 Breaking Strength Results—Yes Indicates Tissue Failure

TRIAL	DISTANCE OF DROP					
	$x_1 = 7$ cm	$x_2 = 8$ cm	$x_3 = 9$ cm	$x_4 = 10$ cm		
1						
2						
3						
4						
5						
6						
7						
8						
9						
10						
11						
12						
13						
14						
15						
16						
Total: yes	$Y_1 =$	$Y_2 =$	$Y_3 =$	$Y_4 =$		
Total: no						

Step 3: Maximum Likelihood Estimation

We will estimate the unknown parameters α and β using maximum likelihood estimation. We must first construct the likelihood function, $L(\alpha, \beta)$. The likelihood function has a factor for each tissue. We have a Bernoulli trial for each tissue. If the weight falls through a tissue, the likelihood factor is $F_{\alpha\beta}(x)$—the probability that a weight dropped from distance x will fall through the tissue. If the weight does not fall through the tissue, the likelihood factor is $1 - F_{\alpha\beta}(x)$.

We can simplify the likelihood function by introducing some additional notation. Let k denote the number of dropping distances used in the experiment. If you dropped weights from 7, 8, 9, and 10 cm, then $k = 4$. If you had to use additional distances, then k will be larger than 4. Let $x_1 = 7$ cm, $x_2 = 8$ cm, $x_3 = 9$ cm, and $x_4 = 10$ cm. If you used additional distances, denote them by x_5, x_6, etc.

Recall that 16 tissues were tested at each dropping distance. Let Y_i denote the number of times that the weight fell through the tissue for dropping distance x_i, $i = 1, \ldots, k$. Thus, if the weight fell through the tissue 3 of the 16 times it was dropped from 7 cm, then $Y_1 = 3$.

The 16 tissues that have dropping distance x_i contribute 16 factors to $L(\alpha, \beta)$. These factors can be summarized as

$$\left[F_{\alpha\beta}(x_i)\right]^{Y_i}\left[1 - F_{\alpha\beta}(x_i)\right]^{16-Y_i} \tag{9.4}$$

Multiplying the contributions to the likelihood function from all k dropping distances yields

$$L(\alpha, \beta) = \prod_{i=1}^{k}\left[F_{\alpha\beta}(x_i)\right]^{Y_i}\left[1 - F_{\alpha\beta}(x_i)\right]^{16-Y_i} \tag{9.5}$$

$$= \prod_{i=1}^{k}\left\{\left[\frac{\exp(\alpha + \beta x_i)}{1 + \exp(\alpha + \beta x_i)}\right]^{Y_i}\left[\frac{1}{1 + \exp(\alpha + \beta x_i)}\right]^{16-Y_i}\right\} \tag{9.6}$$

$$= \prod_{i=1}^{k}\frac{\exp\left[Y_i(\alpha + \beta x_i)\right]}{\left[1 + \exp(\alpha + \beta x_i)\right]^{16}} = \frac{\exp\left[\sum_{i=1}^{k}Y_i(\alpha + \beta x_i)\right]}{\prod_{i=1}^{k}\left[1 + \exp(\alpha + \beta x_i)\right]^{16}} \tag{9.7}$$

The maximum likelihood estimates, $\hat{\alpha}$ and $\hat{\beta}$, are the values of α and β that maximize $L(\alpha, \beta)$ for a given set of data. The requirement that we have at least two distances with both yes and no responses ensures that a unique (α, β) maximizes $L(\alpha, \beta)$.

The logarithm function is an increasing function, so we can find the maximum likelihood estimates by maximizing the log-likelihood function

$$\log[L(\alpha,\beta)] = \sum_{i=1}^{k} Y_i(\alpha + \beta x_i) - 16\sum_{i=1}^{k} \log[1 + \exp(\alpha + \beta x_i)] \tag{9.8}$$

Setting the partial derivatives of $\log[L(\alpha, \beta)]$ with respect to α and β equal to zero yields the following normal equations:

$$\frac{\partial \log[L(\alpha,\beta)]}{\partial \alpha} = \sum_{i=1}^{k} Y_i - 16\sum_{i=1}^{k} \frac{\exp(\alpha + \beta x_i)}{1 + \exp(\alpha + \beta x_i)} = 0 \tag{9.9}$$

$$\frac{\partial \log[L(\alpha,\beta)]}{\partial \beta} = \sum_{i=1}^{k} Y_i x_i - 16\sum_{i=1}^{k} \frac{x_i \exp(\alpha + \beta x_i)}{1 + \exp(\alpha + \beta x_i)} = 0 \tag{9.10}$$

The normal equations do not have closed-form solutions for α and β. It is necessary to make initial guesses—α_0 and β_0—of the solutions and to use numerical techniques to iterate to the solutions.

One simple approach for choosing α_0 and β_0 is based on the fact that

$$\log\left(\frac{F_{\alpha\beta}(x_i)}{1 - F_{\alpha\beta}(x_i)}\right) = \alpha + \beta x_i \tag{9.11}$$

Because $Y_i/16$ is an estimate of $F_{\alpha\beta}(x_i)$, we have

$$\log\left(\frac{\dfrac{Y_i}{16}}{1 - \dfrac{Y_i}{16}}\right) = \log\left(\frac{Y_i}{16 - Y_i}\right) = \log\left(\frac{\text{number of yes responses at } x_i}{\text{number of no responses at } x_i}\right) \tag{9.12}$$

as an estimate of $\alpha + \beta x$, provided that $0 < Y_i < 16$. Setting

$$\log\left(\frac{\text{number of yes responses at } x_i}{\text{number of no responses at } x_i}\right) = \alpha_0 + \beta_0 x_i \tag{9.13}$$

for the two most widely separated x values having both yes and no responses yields two linear equations in two unknowns. Solving these equations for α_0 and β_0 yields the initial guesses.

TABLE 9.2 Example Data for Tissue Experiment

DISTANCE OF DROP	$x_1 = 7$ cm	$x_2 = 8$ cm	$x_3 = 9$ cm	$x_4 = 10$ cm
Number of yes responses	$Y_1 = 2$	$Y_2 = 4$	$Y_3 = 6$	$Y_4 = 13$
Number of no responses	14	12	10	3
Total number of tissues	16	16	16	16

Table 9.2 contains example data for this experiment. The example data may be quite different from yours due to differences in tissue brands. The initial guesses of α and β for the example data are the solutions of

$$-1.94591 = \log\left(\frac{2}{14}\right) = \alpha_0 + 7\beta_0 \text{ and } 1.46634 = \log\left(\frac{13}{3}\right) = \alpha_0 + 10\beta_0 \quad (9.14)$$

Solving Equations (9.14) yields

$$\alpha_0 = -9.90786 \text{ and } \beta_0 = 1.13742 \quad (9.15)$$

as the initial guesses for the example data.

Use Table 9.1 to identify the two most widely separated x's in your data having both yes and no responses and then complete the following equations using the form of Equation (9.13) for these two x values:

$$\underline{\hspace{2cm}} = \log(\underline{\hspace{1cm}} / \underline{\hspace{1cm}}) = \alpha_0 + \beta_0 \underline{\hspace{1cm}} \quad (9.16)$$

$$\underline{\hspace{2cm}} = \log(\underline{\hspace{1cm}} / \underline{\hspace{1cm}}) = \alpha_0 + \beta_0 \underline{\hspace{1cm}} \quad (9.17)$$

Solve these two equations and write the initial guesses for your data:

$$\alpha_0 = \underline{\hspace{2cm}} \text{ and } \beta_0 = \underline{\hspace{2cm}} \quad (9.18)$$

To use Newton's method to find the maximum likelihood estimates, we need the following higher order partial derivatives of $\log[L(\alpha, \beta)]$:

$$\frac{\partial^2 \log[L(\alpha,\beta)]}{\partial\alpha^2} = -16\sum_{i=1}^{k} \frac{\exp(\alpha+\beta x_i)}{[1+\exp(\alpha+\beta x_i)]^2} < 0 \quad (9.19)$$

$$\frac{\partial^2 \log[L(\alpha,\beta)]}{\partial\beta^2} = -16\sum_{i=1}^{k} \frac{x_i^2 \exp(\alpha+\beta x_i)}{[1+\exp(\alpha+\beta x_i)]^2} < 0 \quad (9.20)$$

$$\frac{\partial^2 \log[L(\alpha,\beta)]}{\partial\alpha\partial\beta} = -16\sum_{i=1}^{k} \frac{x_i \exp(\alpha+\beta x_i)}{[1+\exp(\alpha+\beta x_i)]^2} \quad (9.21)$$

With some algebra, we can show that

$$\frac{\partial^2 \log[L(\alpha,\beta)]}{\partial \alpha^2} \frac{\partial^2 \log[L(\alpha,\beta)]}{\partial \beta^2} - \left\{\frac{\partial^2 \log[L(\alpha,\beta)]}{\partial \alpha \partial \beta}\right\}^2 > 0 \qquad \textbf{(9.22)}$$

By the second derivative test, the solution to Equations (9.9) and (9.10) is unique and yields the global maximum for $\log[L(\alpha, \beta)]$; that is, it yields the maximum likelihood estimates.

To maximize $\log[L(\alpha, \beta)]$ using Newton's method, we begin with our initial guesses α_0 and β_0 and then use the iterative equation

$$\begin{pmatrix} \alpha_{k+1} \\ \beta_{k+1} \end{pmatrix} = \begin{pmatrix} \alpha_k \\ \beta_k \end{pmatrix} - \begin{pmatrix} \dfrac{\partial^2 \log[L(\alpha_k,\beta_k)]}{\partial \alpha^2} & \dfrac{\partial^2 \log[L(\alpha_k,\beta_k)]}{\partial \alpha \partial \beta} \\ \dfrac{\partial^2 \log[L(\alpha_k,\beta_k)]}{\partial \alpha \partial \beta} & \dfrac{\partial^2 \log[L(\alpha_k,\beta_k)]}{\partial \beta^2} \end{pmatrix}^{-1} \begin{pmatrix} \dfrac{\partial \log[L(\alpha_k,\beta_k)]}{\partial \alpha} \\ \dfrac{\partial \log[L(\alpha_k,\beta_k)]}{\partial \beta} \end{pmatrix} \qquad \textbf{(9.23)}$$

to find the maximum likelihood estimates. We generally stop the iteration when

$$|\alpha_{k+1} - \alpha_k| \text{ and } |\beta_{k+1} - \beta_k| \qquad \textbf{(9.24)}$$

are sufficiently close to zero; for example, smaller than 0.0000001 when programming in double precision Fortran.

Using the example data in Table 9.2 and the initial guesses in Equations (9.15), Newton's method yields the maximum likelihood estimates

$$\hat{\alpha} = -9.98504 \text{ and } \hat{\beta} = 1.10374 \qquad \textbf{(9.25)}$$

after only four iterations of Equation (9.23). These estimates are quite close to the initial guesses given in Equations (9.15).

Logistic regression is a common statistical technique, so it is not surprising that we can find estimates of α and β using statistical computing packages. The following SAS code produces maximum likelihood estimates for the example data in Table 9.2:

```
data one;
input x n y;
cards;
7 16 2
8 16 4
9 16 6
10 16 13
```

```
proc logistic;
model y/n=x;
run;
```

Step 4: Assessing the Goodness of Fit of the Model

The maximum likelihood estimation process uses the criterion of maximizing the likelihood function to fit the "best" logistic distribution function $F_{\alpha\beta}(x)$ to the data. This fitted model may fit our data quite well. If the underlying distribution of breaking strengths is *not* logistic, it is possible that the "best" logistic model may give a poor fit to our data. Therefore, finding $\hat{\alpha}$ and $\hat{\beta}$ is not the end of our work. We must assess the goodness of fit of this "best" logistic regression model. We will consider two approaches to assess the fit.

The first approach is to compare the sample proportion of tissue failures, $Y_i/16$, and the estimated probability of failure,

$$F_{\hat{\alpha}\hat{\beta}}(x_i) = \frac{\exp(\hat{\alpha}+\hat{\beta}x_i)}{1+\exp(\hat{\alpha}+\hat{\beta}x_i)} \tag{9.26}$$

for each x_i. These values are given in Table 9.3 for the example data.

The observed proportions and the estimated probabilities can also be compared by constructing a multiple scatter plot of both variables versus distance. Figure 9.5 gives a plot for the example data, with asterisks denoting the observed proportions and the solid curve denoting estimated probability. Table 9.3 and Figure 9.5 show that the logistic regression model fits the observed proportions of failures reasonably well, but there are some departures for the two largest distances.

If we were to repeat the experiment, we would likely get different numbers of failures. Thus, we cannot expect the observed proportions of

TABLE 9.3 Observed Proportions of Failure and Estimated Probability of Failure Under the "Best" Logistic Regression Model for the Example Data

DISTANCE x_i	OBSERVED PROPORTION OF FAILURE $Y_i/16$	ESTIMATED PROBABILITY OF FAILURE $F_{\hat{\alpha}\hat{\beta}}(x_i)$
7 cm	2/16 = 0.1250	0.0946
8 cm	4/16 = 0.2500	0.2396
9 cm	6/16 = 0.3750	0.4872
10 cm	13/16 = 0.8125	0.7412

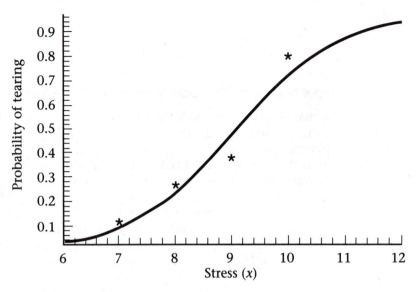

FIGURE 9.5 Comparison of logistic regression model to observed proportions of failures

failures to agree exactly with the estimated probability of failure given by the logistic regression model. But how big do the differences between the observed proportions of failures and the estimated probabilities of failure have to be before we rule out the logistic regression model? To answer this question, we move to our second approach for assessing the goodness of fit of the logistic regression model, the likelihood ratio test for lack of fit.

The maximum likelihood estimators $\hat{\alpha}$ and $\hat{\beta}$ maximize the likelihood function under the restriction that the probability of tissue failure is of the form $F_{\alpha\beta}(x)$ for some unknown values of α and β. Substituting the values of $\hat{\alpha}$ and $\hat{\beta}$ into Equation (9.5) for the example data yields

$$L(\hat{\alpha}, \hat{\beta}) = 1.629050633 \cdot 10^{-15} \tag{9.27}$$

If we remove the restrictions that the probability of tissue failure is of the form $F_{\alpha\beta}(x)$, we have a more general likelihood function

$$L^*(p_1, \ldots, p_k) = \prod_{i=1}^{k} (p_i)^{Y_i} (1 - p_i)^{16 - Y_i} \tag{9.28}$$

for $0 \le p_i \le 1$, $i = 1, \ldots, k$. The maximum of $L^*(p_1, \ldots, p_k)$, which occurs at $\hat{p}_i = Y_i/16$, $i = 1, \ldots, k$, is

$$L^*(\hat{p}_1, \ldots, \hat{p}_k) = \prod_{i=1}^{k} (\hat{p}_i)^{Y_i} (1 - \hat{p}_i)^{16 - Y_i} \tag{9.29}$$

Notice that the more general likelihood function in Equation (9.28) becomes the logistic regression likelihood function in Equation (9.5) if we add the restriction $p_i = F_{\alpha\beta}(x_i)$ for $i = 1, \ldots, k$. The maximization of L^* is over a larger set of possible p_i's than the maximization of L, so we have

$$L(\hat{\alpha}, \hat{\beta}) \leq L^*(\hat{p}_1, \ldots, \hat{p}_k) \tag{9.30}$$

Substituting the Y_i's for the example data from Table 9.2 into Equation (9.29) yields

$$L^*(\hat{p}_1, \ldots, \hat{p}_k) = 3.342899846 \cdot 10^{-15} \tag{9.31}$$

Let $G(x)$ denote the probability of tissue failure as a function of x. The likelihood ratio test of

$$H_0: G(x) = F_{\alpha\beta}(x) \text{ for some value of } \alpha \text{ and } \beta$$
$$H_a: G(x) \neq F_{\alpha\beta}(x) \text{ for any values of } \alpha \text{ and } \beta \tag{9.32}$$

compares $L(\hat{\alpha}, \hat{\beta})$ and $L^*(\hat{p}_1, \ldots, \hat{p}_k)$. If H_0 is true, the two quantities should be relatively close. Under H_a, $L(\hat{\alpha}, \hat{\beta})$ may be considerably smaller than $L^*(\hat{p}_1, \ldots, \hat{p}_k)$.

The null hypothesis is rejected at the 5% significance level if the likelihood ratio statistic

$$-2\log\left[\frac{L(\hat{\alpha}, \hat{\beta})}{L^*(\hat{p}_1, \ldots, \hat{p}_k)}\right] \tag{9.33}$$

is greater than the upper .05 probability point of a chi-square distribution with $k - 2$ degrees of freedom. This is an approximate test based on large-sample theory. The approximation improves as the number of trials at each value of x increases.

For the example data, we would not reject the logistic regression model at the 5% significance level because

$$-2\log\left[\frac{L(\hat{\alpha}, \hat{\beta})}{L^*(\hat{p}_1, \ldots, \hat{p}_k)}\right] = 1.4377 < 5.99$$

Thus, the differences between the observed proportions of failures and the estimated probabilities of failure are consistent with what could have reasonably been expected under the logistic regression model. We can never prove a null hypothesis, thus the likelihood ratio test cannot prove that the logistic regression model is correct. The likelihood ratio test shows that the logistic regression model is not unreasonable for the example data. Other models can also be found that will not be rejected by a likelihood ratio test.

■ ■ ■ ■ ■ PARTING GLANCES

In this engineering experiment we developed a logistic regression model for the probability of tissue failure as a function of the distance the weight is dropped. This analysis is based on the assumption that the breaking strengths of the tissues follow a logistic distribution. Similar analyses can be done for other families of breaking strength distributions by replacing the logistic distribution function $F_{\alpha\beta}(x)$ by the appropriate distribution function. Other common distributions in this type of study are the normal, lognormal, and Weibull.

Logistic regression is also used in medical and psychological experiments. In these experiments, one models the probability that a subject responds when given a fixed stimulus. In medical trials, the stimulus may be a drug dosage given to a patient, and the desired response is a reduction in symptoms. In a psychological experiment, the stimulus might be a tone volume, and the desired response is for the subject to hear the tone.

We considered a simple case of logistic regression with one predictor variable, x. More advanced discussions of logistic regression allow for more than one predictor variable—the logistic regression analog of multiple regression.

■ ■ ■ ■ ■ ASSIGNMENT: ANALYZING THE DATA

Include a copy of Table 9.1 with your responses.

1. Write and run a computer program to find $\hat{\alpha}$ and $\hat{\beta}$ for the data in Table 9.1 using Newton's method and the initial guesses in Equations (9.18).

2. Use a logistic regression routine in a statistical computing package to estimate α and β for the data in Table 9.1. Do the results of the statistical computing package agree with the results of your program using Newton's method?

3. Assess the fit of the logistic regression model for the data in Table 9.1 by creating and studying a table similar to Table 9.3 and a plot similar to Figure 9.5. Based on these results, are you satisfied with the assumption that the breaking strength of the tissues follows a logistic distribution? If not, suggest a more appropriate distribution.

4. Assess the fit of the logistic regression model for the data in Table 9.1 by performing a likelihood ratio test at the 5% significance level. Based on these results, are you satisfied with the assumption that the breaking strength of the tissues follows a logistic distribution? If not, can you suggest a more appropriate distribution?
5. What further analysis, if any, would you suggest for this data? Explain your answer.

■ ■ ■ ■ ■ **ASSIGNMENT: LOOKING AT THE MATHEMATICS**

1. Use the results in Equations (9.19), (9.20), and (9.21) to show that

$$\frac{\partial^2 \log[L(\alpha,\beta)]}{\partial\alpha^2} \frac{\partial^2 \log[L(\alpha,\beta)]}{\partial\beta^2} - \left\{\frac{\partial^2 \log[L(\alpha,\beta)]}{\partial\alpha\partial\beta}\right\}^2 > 0$$

(*Hint*: Use summation indices i and j in the expressions for

$$\frac{\partial^2 \log[L(\alpha,\beta)]}{\partial\alpha^2} \quad \text{and} \quad \frac{\partial^2 \log[L(\alpha,\beta)]}{\partial\beta^2}$$

respectively. Write

$$\left\{\frac{\partial^2 \log[L(\alpha,\beta)]}{\partial\alpha\partial\beta}\right\}^2$$

as the product of two identical factors and use the summation indices i and j in the expressions for the first and second factors, respectively.)
2. Show that no finite (α, β) value would maximize $L(\alpha, \beta)$ if the data consisted of all no responses for $x = 7$ and 8, and all yes responses for $x = 9$ and 10. [*Hint*: Argue that there exists a sequence of (α, β) values such that the sequence of $L(\alpha, \beta)$ values approaches 1 and that $L(\alpha, \beta) < 1$ for all finite (α, β).]

■ ■ ■ ■ ■ **ASSIGNMENT: THE WRITTEN PRESENTATION**

Turn in a copy of Table 9.1 separate from your written presentation.

In your role as a statistician for the Cry Your Eyes Out Tissue Company, write a formal report summarizing your work to be read by process engineer Sheila McDuff, her boss, and your boss. Your report should include the following:

1. A summary of the experiment that was performed and its purpose

2. A description of the model for the probability of tissue failure as a function of the dropping distance of the weight and assumptions on which this model is based
3. Maximum likelihood estimates of the unknown parameters and the probability of tissue failure as a function of the dropping distance of the weight
4. An assessment of the goodness of fit of the model
5. Recommendations regarding the use of the model or the need for further analysis
6. Suggestions for future experiments or improvements to this experiment
7. A technical appendix describing the likelihood function, the numerical techniques used to find the maximum likelihood estimators, and the details of the goodness-of-fit test
8. Professional-quality tables and figures, as necessary

For guidance, refer to Chapter 12: Strategies for Effective Written Reports.

▪ ▪ ▪ ▪ ▪ ASSIGNMENT: THE ORAL PRESENTATION

In your role as a statistician for the Cry Your Eyes Out Tissue Company, prepare a 5-minute oral presentation including professional-quality visual displays to present the results of this experiment to process engineer Sheila McDuff, her boss, and your boss. Your presentation should address items 1–6 and 8 of The Written Presentation assignment section. You should deemphasize the mathematical development in your presentation. For guidance, refer to Chapter 13: Strategies for Effective Oral Presentations.

CHAPTER

10

Estimating Voter Preferences

■ ■ ■ ■ ■ **INTRODUCTION**

Statistics is the science of making decisions in the face of uncertainty. We select a random sample to represent a population and then use the sample results to make decisions about the population. We realize that a different random sample would give us somewhat different sample results. These differences are known as sampling error. Statisticians devise sampling plans and decision rules that maximize the amount of information that can be obtained from sample data.

In this chapter, you will estimate the proportion of individuals in a population of registered voters who support a particular candidate. Our goal will be to choose a sampling plan that makes the standard error of the proportion estimator as small as possible. That is, we want to minimize the uncertainty in the estimate of the proportion of registered voters supporting the candidate.

Computing Concepts and Procedures

Random number generation, SAS PROC SORT

Mathematical Concepts

Minimization subject to an equality constraint

Statistical Concepts

Population proportion, sample-size allocation, sampling error, simple random sample, stratified random sample, standard error

Materials Required

None

■ ■ ■ ■ ■ THE SETTING

You are a statistician employed by the Winners Political Consulting Group. The City of Taxes, Texas, is holding its mayoral election in ten weeks. Lillian White, Taxes' Democratic party chairperson, has hired your company to survey a sample of 100 registered voters to estimate the proportion of registered voters in Taxes who support 72-year-old, Democratic candidate Jane Goodgal. Ms. White explains that three ethnic groups—African American, Caucasian, and Hispanic—are represented in Taxes. She believes, based on past experience, that 75% or more of the African Americans, 30% or less of the Caucasians, and about 50% of the Hispanics will vote for Goodgal. For all ethnic groups, she believes that support for Goodgal will be much stronger for voters over 50 years old than for younger voters, and support may be slightly stronger for females than for males. She gives you a list containing the voter identification number, name, ethnic group, age, and gender of each of the 480 registered voters in Taxes (see Table 10.1). Ms. White will rely on your judgment for the sampling plan. She wants the standard error of the proportion estimator to be as small as possible.

■ ■ ■ ■ ■ THE BACKGROUND

The purpose of this study is to use a random sample of $n = 100$ registered voters to estimate the population proportion

$$P = \frac{\text{number of registered voters supporting Jane Goodgal}}{N} \quad (10.1)$$

where $N = 480$, the number of registered voters in Taxes. The estimator of P and the standard error of the estimator depend on the sampling plan.

A relatively simple sampling method is the simple random sample without replacement. With simple random sampling, all possible samples of 100 registered voters have an equal chance of being selected as our sample. To choose a simple random sample we assign unique identification (ID) numbers to each member of our population.

TABLE 10.1 Registered Voters in Taxes, Texas

ID#	NAME	RACE	AGE	SEX	ID#	NAME	RACE	AGE	SEX
001	Aaron, B.	A	37	F	049	Cepeda, H.	H	34	M
002	Aaron, H.	A	39	M	050	Cepeda, L.	H	33	F
003	Abernathy, C.	C	54	F	051	Conners, A.	A	27	F
004	Abernathy, F.	C	55	M	052	Conners, J.	A	30	M
005	Abernathy, M.	C	82	F	053	Conners, L.	A	50	M
006	Aiken, N.	C	23	M	054	Conners, P.	A	52	F
007	Aiken, O.	C	48	F	055	Cooper, F.	C	47	F
008	Aiken, W.	C	52	M	056	Cooper, M.	C	47	M
009	Aiken, Z.	C	84	F	057	Cooper, P.	C	24	F
010	Arbor, A.	A	21	F	058	Cooper, R.	C	24	M
011	Arbor, B.	A	25	M	059	Cooper, S.	C	49	M
012	Arbor, D.	A	46	F	060	Cooper, W.	C	48	F
013	Arbor, R.	A	46	M	061	Cortez, C.	H	61	M
014	Ayers, J.	C	59	M	062	Cortez, J.	H	60	F
015	Ayers, K.	C	58	F	063	Cortez, O.	H	33	F
016	Ayers, M.	C	27	M	064	Cortez, P.	H	31	M
017	Ayers, P.	C	23	F	065	Cortez, X.	H	26	F
018	Babcock, A.	C	46	M	066	Cutter, D.	C	22	M
019	Babcock, N.	C	46	F	067	Cutter, G.	C	53	F
020	Babcock, P.	C	21	F	068	Dade, D.	C	34	M
021	Babcock, T.	C	22	M	069	Dade, R.	C	31	F
022	Berroa, A.	H	29	M	070	Dade, T.	C	62	F
023	Berroa, F.	H	31	F	071	Dailey, H.	C	54	F
024	Berry, G.	C	57	M	072	Dailey, K.	A	47	M
025	Berry, H.	C	62	F	073	Dailey, L.	C	55	M
026	Berry, L.	C	88	M	074	Davis, B.	C	46	M
027	Berry, P.	C	85	F	075	Davis, C.	C	47	F
028	Bestwick, A.	C	55	F	076	Davis, E.	C	23	M
029	Binford, M.	A	67	M	077	Davis, G.	C	24	F
030	Binford, P.	A	34	M	078	Donovan, A.	A	34	M
031	Binford, Q.	A	31	F	079	Donovan, H.	A	33	F
032	Bonds, G.	A	34	M	080	Donovan, J.	A	56	M
033	Bonds, L.	A	32	F	081	Donovan, L.	A	55	F
034	Bostock, S.	C	46	M	082	Drury, D.	C	44	M
035	Bostock, W.	C	47	F	083	Drury, F.	C	45	F
036	Bostock, X.	C	22	M	084	Drury, P.	C	67	F
037	Byers, B.	C	26	F	085	Drury, W.	C	73	M
038	Byers, R.	C	59	F	086	Druthers, D.	C	67	F
039	Byers, T.	C	61	F	087	Druthers, G.	C	69	M
040	Carson, A.	A	33	F	088	Druthers, S.	C	35	M
041	Carson, B.	A	33	M	089	Druthers, W.	C	30	F
042	Carson, G.	A	57	F	090	Dyer, G.	C	33	F
043	Carson, H.	A	60	M	091	Dyer, K.	C	32	M
044	Castillo, C.	H	24	M	092	Easter, C.	A	45	F
045	Castillo, D.	H	27	F	093	Easter, G.	A	46	M
046	Castillo, H.	H	51	F	094	Easter, N.	A	67	F
047	Castillo, J.	H	52	M	095	Easter, V.	A	23	F
048	Castillo, O.	H	73	F	096	Easter, Z.	A	24	M

(continued on next page)

TABLE 10.1 Registered Voters in Taxes, Texas *(continued)*

ID#	NAME	RACE	AGE	SEX	ID#	NAME	RACE	AGE	SEX
097	Easterling, B.	C	45	F	145	Gomez, A.	H	61	F
098	Easterling, D.	C	47	M	146	Gomez, B.	H	63	M
099	Easterling, F.	C	44	M	147	Gomez, H.	H	38	M
100	Easterling, V.	C	21	M	148	Gomez, L.	H	37	F
101	Easterling, W.	C	22	F	149	Goodgal, J.	C	72	F
102	Engle, B.	C	23	F	150	Gostner, H.	C	43	F
103	Engle, N.	C	25	M	151	Greiner, J.	C	45	M
104	Engle, O.	C	53	F	152	Greiner, P.	C	44	F
105	Engle, P.	C	52	M	153	Greiner, S.	C	18	F
106	Engle, T.	C	79	F	154	Gunter, C.	C	33	M
107	Everett, D.	C	82	F	155	Gunter, S.	C	31	F
108	Everett, S.	C	63	M	156	Gunter, U.	C	44	M
109	Everett, T.	C	59	F	157	Haines, C.	A	44	M
110	Everett, W.	C	34	M	158	Haines, F.	A	42	F
111	Farris, A.	A	45	F	159	Haines, K.	A	20	M
112	Farris, C.	A	44	M	160	Haines, L.	A	48	M
113	Farris, K.	A	67	F	161	Haines, M.	A	66	F
114	Farris, L.	A	21	M	162	Haines, P.	A	68	M
115	Farris, S.	A	23	F	163	Harris, L.	C	87	M
116	Fenton, A.	C	66	F	164	Harris, P.	C	84	F
117	Fenton, N.	C	64	F	165	Harris, R.	C	54	M
118	Fenton, S.	C	66	M	166	Harris, S.	C	49	F
119	Fenton, T.	C	67	M	167	Harrison, A.	C	37	M
120	Fenton, W.	C	39	M	168	Harrison, J.	C	37	F
121	Fenton, Y.	C	37	F	169	Harrison, T.	C	58	M
122	Ferraro, C.	C	19	M	170	Harrison, W.	C	58	F
123	Ferraro, G.	C	21	F	171	Heinz, C.	C	26	M
124	Fortner, B.	C	34	F	172	Heinz, F.	C	25	F
125	Fortner, C.	C	37	M	173	Hermann, D.	C	33	M
126	Fortner, N.	C	60	F	174	Hermann, E.	C	35	M
127	Fortner, P.	C	61	M	175	Hermann, G.	C	34	F
128	Fortner, W.	C	90	F	176	Hermann, J.	C	39	M
129	Foutz, C.	C	83	M	177	Hermann, L.	C	55	F
130	Foutz, D.	C	81	F	178	Hermann, P.	C	57	M
131	Foutz, G.	C	59	M	179	Hermann, Q.	C	61	M
132	Foutz, H.	C	58	F	180	Hermann, W.	C	63	F
133	Gaines, A.	A	67	M	181	Hermann, Y.	C	89	F
134	Gaines, C.	A	66	F	182	Hertz, A.	C	28	M
135	Gaines, D.	A	44	M	183	Hertz, S.	C	27	F
136	Gaines, F.	A	42	F	184	Hertz, T.	C	53	F
137	Gaines, H.	A	40	M	185	Hoover, D.	C	34	M
138	Gaines, K.	A	39	F	186	Hoover, F.	C	33	F
139	Gaines, P.	A	29	F	187	Hoover, L.	C	36	M
140	Gaines, T.	A	31	M	188	Hoover, S.	C	35	F
141	Gaines, W.	A	21	M	189	Hoover, T.	C	60	M
142	Gerry, F.	C	18	M	190	Hunter, L.	A	45	M
143	Gerry, G.	C	25	M	191	Hunter, N.	A	41	F
144	Gerry, L.	C	47	F	192	Hunter, S.	A	19	M

(continued on next page)

TABLE 10.1 Registered Voters in Taxes, Texas *(continued)*

ID#	NAME	RACE	AGE	SEX	ID#	NAME	RACE	AGE	SEX
193	Hunter, T.	A	67	F	241	Jones, P.	C	23	M
194	Ingle, A.	C	65	M	242	Jones, Q.	C	28	F
195	Ingle, C.	C	64	F	243	Jones, W.	C	33	M
196	Ingle, R.	C	41	M	244	Jones, Z.	A	36	F
197	Ingle, S.	C	42	F	245	Juneberry, S.	C	31	F
198	Ingle, T.	C	45	M	246	Kahn, B.	C	33	F
199	Ingle, V.	C	44	F	247	Kahn, C.	C	31	M
200	Ingle, W.	C	21	M	248	Kahn, L.	C	65	F
201	Isley, B.	A	18	M	249	Kahn, P.	C	64	M
202	Isley, H.	A	22	F	250	Kaiser, A.	C	44	M
203	Jacobs, G.	C	25	M	251	Kaiser, N.	C	42	F
204	Jacobs, K.	C	25	F	252	Kaiser, P.	C	21	F
205	Jacobs, M.	C	55	F	253	Kaiser, W.	C	67	M
206	Jacobs, W.	C	54	M	254	Kelly, A.	A	41	M
207	Jarvis, A.	C	52	M	255	Kelly, J.	A	39	F
208	Jarvis, C.	C	49	F	256	Kelly, S.	A	21	M
209	Jarvis, D.	C	24	M	257	Kelly, T.	A	65	F
210	Jarvis, R.	C	23	F	258	Kettinger, S.	C	63	M
211	Jarvis, S.	C	19	M	259	Kettinger, T.	C	62	F
212	Jerald, K.	C	42	M	260	Kettinger, W.	C	37	M
213	Jerald, P.	C	41	F	261	Kettinger, Z.	C	33	F
214	Jerald, R.	C	46	M	262	Koon, B.	C	34	M
215	Jerald, S.	C	39	M	263	Koon, D.	C	39	M
216	Johnson, A.	C	34	M	264	Koon, T.	C	33	F
217	Johnson, B.	C	56	F	265	Kurtz, A.	C	44	M
218	Johnson, D.	C	46	F	266	Kurtz, F.	C	43	F
219	Johnson, F.	C	37	F	267	Kurtz, S.	C	22	M
220	Johnson, G.	A	67	M	268	Lawrence, D.	C	54	M
221	Johnson, J.	A	64	F	269	Lawrence, G.	C	55	F
222	Johnson, L.	A	32	M	270	Lawrence, L.	C	30	M
223	Johnson, M.	A	29	F	271	Lawrence, N.	C	27	F
224	Johnson, S.	C	77	M	272	Lawrence, S.	C	22	F
225	Johnson, T.	C	73	F	273	Laws, B.	C	34	M
226	Johnson, U.	C	52	F	274	Laws, F.	C	32	F
227	Johnson, W.	C	39	F	275	Lee, A.	A	45	M
228	Johnson, Z.	C	24	M	276	Lee, D.	A	37	F
229	Jones, A.	A	34	M	277	Lee, F.	A	23	M
230	Jones, B.	A	32	F	278	Lee, N.	C	56	M
231	Jones, C.	A	45	M	279	Lee, P.	C	54	F
232	Jones, E.	A	62	F	280	Leemis, D.	C	55	F
233	Jones, F.	A	76	M	281	Leemis, G.	C	55	M
234	Jones, G.	C	54	M	282	Lergins, H.	C	40	F
235	Jones, H.	C	51	F	283	Lewis, A.	C	42	M
236	Jones, K.	C	32	M	284	Lewis, N.	C	44	F
237	Jones, L.	C	34	F	285	Lewis, P.	C	29	M
238	Jones, M.	C	56	M	286	Lewis, T.	C	31	F
239	Jones, N.	C	53	M	287	Lloyd, D.	C	33	M
240	Jones, O.	C	67	F	288	Lloyd, G.	C	31	F

(continued on next page)

TABLE 10.1   Registered Voters in Taxes, Texas *(continued)*

ID#	NAME	RACE	AGE	SEX	ID#	NAME	RACE	AGE	SEX
289	Lloyd, T.	C	67	F	337	Nunez, H.	H	64	F
290	Loadholt, S.	C	77	M	338	Nunez, L.	H	39	M
291	Loadholt, T.	C	73	F	339	Nunez, P.	H	34	F
292	Lobard, W.	C	56	F	340	Oates, J.	C	43	M
293	Louis, A.	C	67	M	341	Oates, K.	C	42	F
294	Louis, C.	C	66	F	342	Oates, T.	C	19	M
295	Louis, N.	C	43	M	343	Oberlee, A.	C	23	F
296	Louis, R.	C	48	F	344	Oberlee, N.	C	24	M
297	Louis, T.	C	40	F	345	Oberlee, P.	C	49	F
298	Louis, W.	C	24	M	346	O'Cain, M.	C	42	M
299	Luther, C.	C	29	M	347	O'Cain, S.	C	40	F
300	Luther, G.	C	28	F	348	Olsen, B.	C	34	M
301	Luther, J.	C	33	M	349	Olsen, F.	C	32	F
302	Luther, L.	C	36	F	350	Overton, D.	C	45	M
303	Maas, G.	C	49	M	351	Overson, F.	C	44	F
304	Maas, R.	C	47	F	352	Overton, S.	C	22	M
305	Maas, T.	C	19	M	353	Overton, T.	C	19	F
306	Maise, P.	C	66	M	354	Perez, B.	H	18	M
307	Maise, W.	C	65	F	355	Perez, F.	H	21	F
308	Maryland, D.	A	44	M	356	Perez, L.	H	46	M
309	Maryland, K.	A	42	F	357	Perez, M.	H	45	F
310	Maryland, P.	A	41	M	358	Perez, P.	H	20	M
311	Maryland, T.	A	39	F	359	Perry, D.	A	23	M
312	Mason, B.	C	56	M	360	Perry, S.	A	46	F
313	Mason, H.	C	53	F	361	Perry, W.	A	18	F
314	Mason, T.	C	50	F	362	Porter, A.	C	43	M
315	Mason, U.	C	49	M	363	Porter, N.	C	42	F
316	Mattis, S.	C	78	M	364	Porter, T.	C	19	M
317	Mattis, T.	C	73	F	365	Quattlebaum, D.	C	34	M
318	Moe, S.	C	50	M	366	Quattlebaum, R.	C	33	F
319	Moe, T.	C	26	F	367	Randles, D.	C	38	M
320	Morris, D.	C	32	M	368	Randles, H.	C	37	F
321	Morris, G.	C	29	F	369	Rice, K.	C	34	M
322	Morris, K.	C	55	M	370	Rice, N.	C	31	F
323	Morris, N.	C	54	F	371	Rice, P.	C	68	M
324	Morris, P.	C	20	F	372	Rooney, F.	C	47	M
325	Morrison, D.	C	28	M	373	Rooney, J.	C	44	F
326	Morrison, R.	C	26	F	374	Rooney, S.	C	19	F
327	Murphy, E.	A	42	M	375	Rooney, T.	C	22	F
328	Murphy, G.	A	40	F	376	Sanders, D.	C	34	M
329	Murphy, L.	A	18	F	377	Sanders, F.	A	45	F
330	Murphy, S.	C	34	M	378	Sanders, H.	A	43	M
331	Murphy, T.	C	32	F	379	Sanders, P.	C	31	F
332	Nameth, C.	C	34	M	380	Saunders, M.	C	34	M
333	Nameth, S.	C	33	F	381	Saunders, P.	C	35	F
334	Norris, D.	C	40	M	382	Sears, L.	C	27	M
335	Norris, M.	C	38	F	383	Sears, P.	C	28	F
336	Nunez, B.	H	66	M	384	Seivers, H.	C	39	F

(continued on next page)

TABLE 10.1 Registered Voters in Taxes, Texas *(continued)*

ID#	NAME	RACE	AGE	SEX	ID#	NAME	RACE	AGE	SEX
385	Seivers, P.	C	39	M	433	Walters, F.	A	36	F
386	Simmons, G.	C	45	M	434	Walters, G.	A	38	M
387	Simmons, K.	C	43	F	435	Walters, R.	A	62	M
388	Smith, A.	C	48	M	436	Walters, T.	A	58	F
389	Smith, N.	C	47	F	437	Waters, B.	C	48	M
390	Smith, P.	C	23	M	438	Waters, G.	C	47	M
391	Smith, T.	C	18	F	439	Waters, R.	C	22	M
392	Southern, B.	C	26	M	440	Waters, T.	C	19	F
393	Southern, L.	C	23	F	441	Wilson, F.	C	24	M
394	Southern, T.	C	60	F	442	Wilson, J.	C	29	F
395	Suggs, G.	C	22	M	443	Wilson, K.	C	32	F
396	Talley, H.	A	25	M	444	Wilson, R.	C	35	M
397	Talley, L.	A	31	F	445	Wilson, S.	C	49	M
398	Talley, P.	A	55	F	446	Wilson, T.	C	54	F
399	Talley, X.	A	57	M	447	Wilson, W.	C	55	M
400	Taylor, G.	C	34	M	448	Woodrow, D.	C	47	M
401	Taylor, P.	C	32	F	449	Woodrow, M.	C	45	F
402	Taylor, R.	C	45	F	450	Woodrow, N.	C	23	M
403	Taylor, T.	C	47	M	451	Woodrow, P.	C	19	F
404	Thomas, A.	C	35	M	452	Woods, E.	C	24	M
405	Thomas, C.	C	37	F	453	Woods, G.	C	22	F
406	Thomas, F.	C	61	M	454	Woods, H.	C	50	F
407	Thomas, K.	C	59	F	455	Woods, L.	C	52	M
408	Thompson, R.	C	23	M	456	Woods, P.	C	18	F
409	Thompson, S.	C	25	M	457	Woods, W.	C	29	M
410	Thompson, T.	C	19	F	458	Wonders, S.	A	32	F
411	Thompson, W.	C	50	F	459	Wonders, T.	A	34	M
412	Torrez, B.	H	34	M	460	Wormier, A.	C	31	M
413	Torrez, R.	H	32	F	461	Wormier, J.	C	29	F
414	Turner, B.	A	54	M	462	Yonce, A.	C	23	M
415	Turner, D.	A	52	F	463	Yonce, B.	C	19	F
416	Turner, G.	A	30	M	464	Yonce, F.	C	45	F
417	Turner, K.	A	29	F	465	Yonce, M.	C	55	M
418	Turner, S.	A	33	M	466	Young, C.	C	42	M
419	Ulmer, A.	C	42	M	467	Young, D.	C	37	F
420	Ulmer, B.	C	41	M	468	Young, K.	C	66	M
421	Ulmer, G.	C	39	F	469	Young, P.	C	65	F
422	Ulmer, R.	C	37	F	470	Young, Y.	C	32	M
423	Ulmer, V.	C	18	F	471	Young, X.	C	34	F
424	Utter, A.	C	28	M	472	Zahn, C.	C	23	M
425	Utter, G.	C	26	F	473	Zahn, G.	C	26	F
426	Vail, D.	C	33	M	474	Zahn, L.	C	49	M
427	Vail, T.	C	32	F	475	Zahn, P.	C	47	F
428	Vargo, G.	C	32	M	476	Zellmer, B.	C	56	M
429	Vargo, L.	C	29	F	477	Zellmer, C.	C	23	F
430	Vasquez, B.	H	33	M	478	Zellmer, G.	C	49	F
431	Vasquez, L.	H	32	F	479	Zellmer, P.	C	19	M
432	Vasquez, T.	H	59	F	480	Zellmer, W.	C	28	F

We then use a computer or a random number table to generate a random stream of these ID numbers. Our simple random sample will be the 100 individuals whose ID numbers appear first in our stream of random numbers. For example, the registered voters are numbered from 001 to 480 in Table 10.1. The following stream of random integers between 001 and 480 was generated using MINITAB:

096, 201, 465, 068, 167, . . .

Using these random numbers, the first 5 individuals in the simple random sample of voters would be: Z. Easter, B. Isley, M. Yonce, D. Dade, and A. Harrison. When sampling without replacement, duplicate occurrences of random numbers are ignored.

The estimator of P for a simple random sample of n voters is

$$\hat{P} = \frac{\text{number of voters in sample supporting Jane Goodgal}}{n} \tag{10.2}$$

The standard error of the simple random sample estimator is

$$\left\{ \frac{[P(1-P)]}{n} \frac{(N-n)}{(N-1)} \right\}^{1/2} \tag{10.3}$$

Because P is unknown, we can use the standard error estimate

$$\left\{ \frac{[\hat{P}(1-\hat{P})]}{(n-1)} \frac{(N-n)}{N} \right\}^{1/2} \tag{10.4}$$

An approximate 95% confidence interval for P is

$$\hat{P} \pm 2 \left\{ \frac{[\hat{P}(1-\hat{P})]}{(n-1)} \frac{(N-n)}{N} \right\}^{1/2} \tag{10.5}$$

A somewhat more complicated sampling method is known as stratified random sampling without replacement. In stratified random sampling, we partition the population into subsets, called strata. Each member of the population is in one and only one stratum. Let L denote the number of strata. We then select a simple random sample of size n_i without replacement from the N_i members of stratum i for each of $i = 1, \ldots, L$.

The true proportion of registered voters in stratum i supporting Goodgal is

$$P_i = \frac{\text{number of registered voters in stratum } i \text{ supporting Goodgal}}{N_i} \tag{10.6}$$

We use the sample items from stratum i to estimate P_i with the estimator

$$\hat{P}_i = \frac{\text{number of voters in sample from stratum } i \text{ supporting Goodgal}}{n_i} \quad \textbf{(10.7)}$$

The stratified random sample estimator of the population proportion, P, is the weighted sum

$$\hat{P}_{st} = \sum_{i=1}^{L} \frac{N_i \hat{P}_i}{N} \quad \textbf{(10.8)}$$

The standard error of the stratified random sample estimator is

$$\left\{ \sum_{i=1}^{L} \frac{N_i^2}{N^2} \frac{[P_i(1-P_i)]}{n_i} \frac{(N_i - n_i)}{(N_i - 1)} \right\}^{1/2} \quad \textbf{(10.9)}$$

The P_i's are unknown, so we use the standard error estimate

$$\left\{ \sum_{i=1}^{L} \frac{N_i^2}{N^2} \frac{\left[\hat{P}_i\left(1-\hat{P}_i\right)\right]}{(n_i - 1)} \frac{(N_i - n_i)}{N_i} \right\}^{1/2} \quad \textbf{(10.10)}$$

An approximate 95% confidence interval for P is given by

$$\hat{P} \pm 2 \left\{ \sum_{i=1}^{L} \frac{N_i^2}{N^2} \frac{\left[\hat{P}_i\left(1-\hat{P}_i\right)\right]}{(n_i - 1)} \frac{(N_i - n_i)}{N_i} \right\}^{1/2} \quad \textbf{(10.11)}$$

The estimated standard error and the approximate confidence interval formulas require that $n_i \geq 2$ for $i = 1, \ldots, L$.

The standard error of the stratified random sample proportion estimator in Equation (10.9) depends on how the population is stratified—the unknown P_i's and the n_i's. The statistician often has considerable flexibility in forming the strata and in allocating the observations among the strata. For fixed values of L and the P_i's, the optimal allocation of observations to the strata uses

$$n_i \approx \frac{n\left[N_i^3 P_i(1-P_i)/(N_i - 1)\right]^{1/2}}{\sum_{k=1}^{L}\left[N_k^3 P_k(1-P_k)/(N_k - 1)\right]^{1/2}} \quad \textbf{(10.12)}$$

where n is the total sample size. The n_i's must be integers satisfying the constraints $2 \leq n_i \leq N_i$, $i = 1, \ldots, L$. For our sample, $n = 100$.

■ ■ ■ ■ ■ **THE EXPERIMENT**

Step 1: The Basic Measurement

The basic measurement will be a response of "Yes," "No," or "Uncertain" to the question "Do you support Jane Goodgal in the upcoming Taxes Mayoral Election?" for each registered voter in the sample. Your instructor will give you the response for each voter in your sample.

Step 2: The Sampling Plan

You are now ready to determine the sampling method and the sample-size allocation. Use a total of 100 observations. The goal is to provide point and interval estimates of the proportion of voters who support Goodgal *such that the standard error of the point estimator is minimized.*

Your student intern, Wannabe A. Statman, has organized the population of voters into 12 possible strata based on ethnicity, gender, and age. The ID numbers for the voters in each strata are given in Table 10.2. Wannabe has not taken a course in sampling and is uncertain whether this is the best way to stratify this population.

Review the information in the section The Setting as you develop your sampling plan. Lillian White has given you some very useful information. However, she has not given you enough information to determine with certainty that a particular sampling plan is the best. You may find this frustrating. It is also the reality of working in the real world—a place with no answers in the back of the book.

Use the information from Ms. White to make "best guesses" of the true proportions favoring Goodgal for the 12 strata in Table 10.2. For example, you might guess, based on Ms. White's information, that 80% of the African Americans support Ms. Goodgal. Her support is much weaker for younger voters, so you might guess that 70% of African Americans under the age of 50 support Ms. Goodgal. You will also need to make a small adjustment for gender.

In general, it is a good idea to combine strata if there is little difference among the P_i's. For any particular stratification, you should use optimal allocation to allocate the total sample size of 100 among the various strata with the restriction that all $n_i \geq 2$. Try various stratification strategies and pick the strategy that minimizes the standard error of the population proportion estimator under the assumption that your "best guesses" are correct. Examples of stratification strategies include the following:

Using the 12 strata in Table 10.2
Combining all 12 strata into 1—simple random sampling
Using three ethnic-based strata—African American, Caucasian, Hispanic
Using gender-based strata—female, male

There are many other possible stratification strategies, some of which may be better than the listed strategies. You may want to use a computer to evaluate the standard errors for various strategies.

TABLE 10.2 Identification Numbers of Registered Voters of Taxes, Texas, Stratified by Ethnicity, Gender, and Age

African American females who are no more than 50 years old (34 individuals)

001	010	012	031	033	040	051	079	092	095
111	115	136	138	139	158	191	202	223	230
244	255	276	309	311	328	329	360	361	377
397	417	433	458						

African American females who are more than 50 years old (14 individuals)

042	054	081	094	113	134	161	193	221	232
257	398	415	436						

African American males who are no more than 50 years old (41 individuals)

002	011	013	030	032	041	052	053	072	078
093	096	112	114	135	137	140	141	157	159
160	190	192	201	222	229	231	254	256	275
277	308	310	327	359	378	396	416	418	434
459									

African American males who are more than 50 years old (10 individuals)

029	043	080	133	162	220	233	399	414	435

Caucasian females who are no more than 50 years old (122 individuals)

007	017	019	020	035	037	055	057	060	069
075	077	083	089	090	097	101	102	121	123
124	144	150	152	153	155	166	168	172	175
183	186	188	197	199	204	208	210	213	218
219	227	237	242	245	246	251	252	261	264
266	271	272	274	282	284	286	288	296	297
300	302	304	314	319	321	324	326	331	333
335	341	343	345	347	349	351	353	363	366
368	370	373	374	375	379	381	383	384	387
389	391	393	401	402	405	410	411	421	422
423	425	427	429	440	442	443	449	451	453
454	456	461	463	464	467	471	473	475	477
478	480								

Caucasian females who are more than 50 years old (55 individuals)

003	005	009	015	025	027	028	038	039	067
070	071	084	086	104	106	107	109	116	117
126	128	130	132	149	164	170	177	180	181
184	195	205	217	225	226	235	240	248	259
269	279	280	289	291	292	294	307	313	317
323	394	407	446	469					

(continued on next page)

TABLE 10.2 Identification Numbers of Registered Voters of Taxes, Texas, Stratified by Ethnicity, Gender, and Age *(continued)*

Caucasian males who are no more than 50 years old (125 individuals)

006	016	018	021	034	036	056	058	059	066
068	074	076	082	088	091	098	099	100	103
110	120	122	125	142	143	151	154	156	167
171	173	174	176	182	185	187	196	198	200
203	209	211	212	214	215	216	228	236	241
243	247	250	260	262	263	265	267	270	273
283	285	287	295	298	299	301	303	305	315
318	320	325	330	332	334	340	342	344	346
348	350	352	362	364	365	367	369	372	376
380	382	385	386	388	390	392	395	400	403
404	408	409	419	420	424	426	428	437	438
439	441	444	445	448	450	452	457	460	462
466	470	472	474	479					

Caucasian males who are more than 50 years old (47 individuals)

004	008	014	024	026	073	085	087	105	108
118	119	127	129	131	163	165	169	178	179
189	194	206	207	224	234	238	239	249	253
258	268	278	281	290	293	306	312	316	322
371	406	447	455	465	468	476			

Hispanic females who are no more than 50 years old (11 individuals)

023	045	050	063	065	148	339	355	357	413
431									

Hispanic females who are more than 50 years old (6 individuals)

046	048	062	145	337	432

Hispanic males who are no more than 50 years old (11 individuals)

022	044	049	064	147	338	354	356	358	412
430									

Hispanic males who are more than 50 years old (4 individuals)

047	061	146	336

Step 3: Sample Selection

After deciding on a sampling plan, describe your plan in Table 10.3. You are now ready to select your sample. You need to carefully document how the sample was taken in the Interim Report section. A description similar to the following will provide sufficient documentation if the 12 strata listed in Table 10.2 are used.

I numbered the N_i items within each stratum in Table 10.2 from 01 to N_i moving across the page from left to right. Using the random number table on pages 426 and 427 of Scheaffer, Mendenhall, and Ott's book *Elementary Survey Sampling* (5th edition), I selected line 21 and column 6 on page 426 as my random starting point. I used the first two digits of the five-digit numbers for strata having fewer than 100 items and the first three digits for strata having at least 100 items. I moved down the page, skipping numbers outside the range of 01 to N_i and duplicate numbers within the same stratum. Upon reaching the bottom of a column, I moved to the top of the next column on the right. Upon coming to the bottom of the last column on page 427, I returned to the top of the first column on page 426. Upon completing the sample selection for a stratum, I began the sample selection for the next stratum using the next random number in the random number table.

If you use a random number table, your documentation should include a photocopy of the table. If you generate random numbers on the computer, your documentation should include a list of all random numbers considered.

Use random numbers to select a sample according to your plan. Write the voter identification numbers of the 100 sampled voters in the order that they were selected on the blank lines at the bottom of Table 10.3. **Important:** The numbers written at the bottom of Table 10.3 must be the voter identification numbers given in Tables 10.1 and 10.2. Do not write other identification numbers that might have been assigned as part of the sample selection process.

■ ■ ■ ■ ■ ASSIGNMENT: INTERIM REPORT

Give your instructor a completed copy of Table 10.3 and documentation of how the sample was selected. The documentation must be detailed enough so that the instructor can follow your instructions and verify that you have correctly selected the sample according to your sampling plan.

The Data

Your instructor will circle the identification numbers in Table 10.3 for all voters supporting Goodgal.

■ ■ ■ ■ ■ PARTING GLANCES

Samples of voter preferences yield important information to political candidates. The candidates can assess how popular they are among various groups. Many candidates will adjust their campaigns depending on the outcomes of polls. During the course of a major campaign, there will be several polls. The proportion of voters planning to support a particular candidate will generally vary in the different polls. Part of this variation is due to sampling error. Different samples of voters give somewhat

TABLE 10.3 Sampling Plan and Selected Sample

Your name _____

_____ I will use a simple random sample.

_____ I will use a stratified random sample using $L =$ _____ strata ($2 \leq L \leq 12$):

STRATA	n_i	n_i	DESCRIPTION (E.G., HISPANIC MALES OVER 50)
1			
2			
3			
4			
5			
6			
7			
8			
9			
10			
11			
12			

Identification numbers of 100 voters in the sample in the order they were selected:

different results. Another source of variation is the dynamic nature of public opinion. Many voters change their mind about who they will vote for during the course of the campaign. To evaluate the results of any poll, we need to know what the population was, how the sample was taken, the uncertainty of the estimator, and the date the poll was taken. The media generally report the uncertainty of the estimator as plus or minus 2 estimated standard errors.

Sample surveys are used in many other areas. Public opinion polls, health surveys, and consumer preference surveys are examples.

Statisticians frequently sort data sets based on values of one or more variables. The following SAS code sorts the data set in Table 10.1 into the 12 strata in Table 10.2:

```
data one;
input id race $ age sex $;
over50 = 'n';
if age > 50 then over50 = 'y';
cards;
001 a 37 f
002 a 39 m
```

An entry for each registered voter goes here.

```
480 c 28 f
proc sort; by race sex over50;
proc print; var id; by race sex over50;
run;
```

■ ■ ■ ■ ■ REFERENCE

Scheaffer, Richard L., Mendenhall, William III, & Ott, R. Lyman (1996), *Elementary Survey Sampling* (5th ed.). Belmont, CA: Duxbury.

■ ■ ■ ■ ■ ASSIGNMENT: ANALYZING THE DATA

Include a copy of Table 10.3 with your responses.

1. Compute the point estimate of the proportion of voters supporting Jane Goodgal that is appropriate for your sampling plan.
2. Compute the estimated standard error and an approximate 95% confidence interval for the proportion of voters supporting Jane Goodgal.

■ ■ ■ ■ ■ ASSIGNMENT: LOOKING AT THE MATHEMATICS

1. Use the fact that a stratified random sample is a collection of simple random samples along with formula (10.3) and equation (10.8) to derive formula (10.9).
2. Assume that there are $L = 2$ strata. Then $n_2 = 100 - n_1$. Use calculus to find the value of n_1 that minimizes the standard error of the stratified random sample estimator, formula (10.9).

■ ■ ■ ■ ■ ASSIGNMENT: THE WRITTEN PRESENTATION

Turn in a copy of Table 10.3 separate from your written presentation.

In your role as a statistician for the Winners Political Consulting Group, write a formal report summarizing your work to be read by Democratic party chairperson Lillian White and your boss, John Jefferson. Your report should include the following:

1. A concisely written descriptive title for the project
2. Your name and the name of your company—Winners Political Consulting Group
3. An executive summary summarizing the purpose and major results of the study in at most four sentences
4. An introduction describing the purpose of the study, including a description of the population and for whom the study was done. The population description should include a breakdown by ethnicity, gender, and age.
5. A methods section giving a precise description of the sampling plan used, including a clear definition of any stratification that was done. All formulas go in the technical appendix (see number 7).
6. A results section that gives a point estimate and a confidence interval for the proportion of voters who support Jane Goodgal. These estimates should be presented in language that can be understood by an educated person who has never studied statistics. If stratified sampling is used, you should also present separate estimates for important subpopulations. Do not include the responses of individual voters.
7. A technical appendix that provides all formulas used to compute point and interval estimates. Carefully define all terms. Your instructor may allow you to write formulas by hand in this report. In a professional report, these would be typed.
8. Professional-quality tables and figures, as necessary.

For guidance, refer to Chapter 12: Strategies for Effective Written Reports.

■ ■ ■ ■ ■ ASSIGNMENT: THE ORAL PRESENTATION

In your role as a statistician for the Winners Political Consulting Group, prepare a 5-minute oral presentation including professional-quality visual displays to present the results of this experiment to Democratic party chairperson Lillian White and your boss, John Jefferson. Your presentation should address items 1–6 and 8 of The Written Presentation assignment section. For guidance, refer to Chapter 13: Strategies for Effective Oral Presentations.

CHAPTER

11

Estimating the Probability
of a Hit in Baseball

■ ■ ■ ■ ■ **INTRODUCTION**

There are several methods for estimating an unknown parameter. Some methods are based entirely on the observed data. These methods, known as frequentist methods, include maximum likelihood, method of moments, and minimum variance unbiased estimators. Other methods, known as Bayesian estimators, are based on both the observed data and on prior information or belief about the unknown parameter.

In this chapter, the unknown parameter is the probability that a particular baseball player will get a base hit against a major-league pitcher. You will be supplied with some prior information about his batting ability and then be asked to determine a Bayesian estimator of the unknown probability. You will compare the Bayesian estimator to frequentist methods for estimating this probability. The estimators will be compared based on their mean squared error functions.

Computing Concepts and Procedures

Two-dimensional plot

Mathematical Concepts

Integration, nonlinear inequality

Statistical Concepts

Bias, Bayesian estimation, Bernoulli trial, binomial distribution, beta distribution, maximum likelihood estimation, mean squared error, method of moments, minimum variance unbiased estimation, squared error loss

Materials Required

None

■ ■ ■ ■ ■ THE SETTING

You are a statistician employed by On The Ball Consulting. Veteran major-league baseball scout Rocky Chew seeks your advice regarding estimating the probability that amateur baseball player John D. Spurrier will get a base hit against a major-league pitcher. Typically, this would be estimated by counting the number of base hits, X, in n "at bats" against the pitcher and reporting the proportion X/n. In the language of baseball, this proportion is the batting average. Rocky has substantial prior information about Spurrier's batting ability and about professional baseball. He thinks the estimator should reflect this information. He has arranged for Spurrier to have ten at bats against a major-league pitcher.

■ ■ ■ ■ ■ THE BACKGROUND

The traditional batting average, $\hat{p}_f = X/n$, is a frequentist estimator in that it makes use of the observed data but ignores any prior information that might exist. If we assume that the at bats are independent Bernoulli trials with a constant probability of getting a base hit, then X has a binomial distribution with parameters n and $p = P$(getting a base hit). Under these assumptions, the batting average is the maximum likelihood estimator, the method of moments estimator, and the minimum variance unbiased estimator of the unknown probability. You might question the assumptions of independence and constant probability of getting a base hit, but they will serve as an initial model for the investigation.

Rocky gives you the following prior information:

■ John D. Spurrier appears to be a good but not great player. He is one of the better batters on a somewhat above-average American Legion baseball team.
■ The few major-league scouts who have watched him play do not believe that Spurrier's batting ability is at the professional level.
■ A barely adequate major-league hitter has a batting average of about 0.200.

- A very good major-league batter has a batting average of about 0.300.
- Ty Cobb had the all-time best major-league career batting average of 0.366.

Bayesian estimation incorporates prior information through a prior density function, $g(p)$, on the unknown parameter p. That is, p is assumed to be the value of a random variable with density function $g(p)$. Because p is a probability, we know that $0 \leq p \leq 1$. Rocky's prior information suggests that the prior density function should put a large amount of density in the $0 \leq p \leq 0.2$ interval and very little density above $p = 0.300$.

A popular family of prior density functions for the binomial parameter p is the beta family. This family depends on two positive parameters, α and β. The density function is

$$g(p) = \frac{\Gamma(\alpha + \beta)}{\Gamma(\alpha)\Gamma(\beta)} p^{\alpha-1}(1 - p)^{\beta-1} \qquad \text{for } 0 \leq p \leq 1 \qquad (11.1)$$

where Γ denotes the gamma function. Recall that $\Gamma(\alpha) = (\alpha - 1)!$ if α is a positive integer. The mean and variance of a beta distribution are $\alpha/(\alpha + \beta)$ and $\alpha\beta/[(\alpha + \beta + 1)(\alpha + \beta)^2]$, respectively.

Different values of α and β give dramatically different density functions. Note that $(\alpha, \beta) = (1, 1)$ gives the uniform density function on $[0,1]$. If $\alpha = \beta$, $g(p)$ is symmetric about $p = 0.5$. If $\alpha < \beta$, $g(p)$ is right skewed. If $\alpha > \beta$, $g(p)$ is left skewed. Figure 11.1 shows $g(p)$ for $(\alpha, \beta) = (2, 2)$, $(2, 5)$, and $(1, 5)$. Rocky's prior information suggests that we choose $\alpha < \beta$. The density functions in Figure 11.1 put more density on larger values of p than would be justified from Rocky's information.

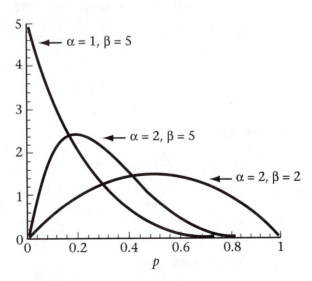

FIGURE 11.1 Beta density functions

Under the Bayesian approach, we consider the joint distribution of X and p. The conditional distribution of X given p is binomial (n, p) and p has a beta prior distribution, so the joint density of X and p is:

$$f(x, p) = \frac{n!}{x!(n-x)!} p^x (1-p)^{n-x} \frac{\Gamma(\alpha+\beta)}{\Gamma(\alpha)\Gamma(\beta)} p^{\alpha-1}(1-p)^{\beta-1}$$

$$= \frac{n!}{x!(n-x)!} \frac{\Gamma(\alpha+\beta)}{\Gamma(\alpha)\Gamma(\beta)} p^{x+\alpha-1}(1-p)^{n-x+\beta-1} \quad \text{for } x = 0, 1, \ldots, n \quad \textbf{(11.2)}$$
$$\text{and } 0 \le p \le 1$$

The marginal distribution of X is found by integrating the joint density over the range of p,

$$h(x) = \int_0^1 f(x, p)dp = \frac{n!}{x!(n-x)!} \frac{\Gamma(\alpha+\beta)}{\Gamma(\alpha)\Gamma(\beta)} \frac{\Gamma(x+\alpha)\Gamma(n-x+\beta)}{\Gamma(n+\alpha+\beta)} \quad \textbf{(11.3)}$$
$$\text{for } x = 0, 1, \ldots, n$$

The posterior density function of p is

$$g(p|X = x) = \frac{f(x, p)}{h(x)} = \frac{\Gamma(n+\alpha+\beta)}{\Gamma(x+\alpha)\Gamma(n-x+\beta)} p^{x+\alpha-1}(1-p)^{n-x+\beta-1} \quad \textbf{(11.4)}$$
$$\text{for } 0 \le p \le 1$$

The posterior distribution of p given $X = x$ is the beta distribution with parameters $x + \alpha$ and $n - x + \beta$.

If we measure the goodness of an estimate $\hat{p}$ by squared error loss, $(\hat{p} - p)^2$, then the Bayesian estimator is the expected value of the posterior distribution. Thus, the Bayesian estimator is

$$\hat{p}_b = \frac{x + \alpha}{n + \alpha + \beta} \quad \textbf{(11.5)}$$

■ ■ ■ ■ ■ THE EXPERIMENT

Step 1: The Basic Measurement

The basic measurement is determining whether the batter gets a base hit in an at bat. *Official Baseball Rules* (1998) gives operational definitions of base hit and at bat.

Step 2: Data Collection

John D. Spurrier will have $n = 10$ at bats against a major-league pitcher. The random variable X will be the number of base hits that he gets.

Step 3: Determination of the Prior Probability Distribution

We have derived the Bayesian estimator based on the beta prior except for the choices of α and β. The choice of α and β is subjective. You should choose α and β such that the prior density function $g(p)$ best matches your understanding of Rocky Chew's prior information. You will want to graph $g(p)$ for several choices of α and β before making your final choice. The following Maple V command produces the plot of the beta density function with $\alpha = 2$ and $\beta = 5$:

```
plot(30*p*(1-p)^4,p=0..1);
```

Remember that we want $\alpha < \beta$ with high density for the $0 \le p \le 0.2$ interval and low density for $p > 0.300$.

After you have graphed several beta density functions, write your choices for α and β:

$\alpha =$ _____ and $\beta =$ _____

Step 4: Comparison of the Estimators

Now let's compare the frequentist estimator $\hat{p}_f = X/n$ and your Bayesian estimator

$$\hat{p}_b = \frac{x+\alpha}{n+\alpha+\beta}$$

We begin by computing both estimators for the possible values of X. The values of $\hat{p}_f$ are given in Table 11.1. Evaluate your Bayesian estimator for $X = 0, 1, \ldots, 10$ and write the results in Table 11.1.

The two estimators give different estimates. The Bayesian estimator is a weighted average of $\hat{p}_f$ and $\alpha/(\alpha + \beta)$, the mean of the prior distribution. Thus, the Bayesian estimator modifies the frequentist estimator by pulling it toward the mean of the prior distribution.

The goodness of an estimator $\hat{p}$ is often measured by its mean squared error,

$$\text{MSE}(\hat{p}) = E[(\hat{p} - p)^2] = \text{Var}(\hat{p}) + [E(\hat{p}) - p]^2 \qquad (11.6)$$

That is, the mean squared error equals the variance plus the square of the bias. For unbiased estimators, the second term is zero and the mean squared error equals the variance. We wish the mean squared error to be as small as possible.

Under our assumptions, X has a binomial distribution with parameters 10 and p. Thus,

$$E(X) = 10p \quad \text{and} \quad \text{Var}(X) = 10p(1-p) \qquad (11.7)$$

TABLE 11.1 Frequentist and Bayesian Estimates of p for $X = 0, 1, \ldots, 10$

X	$\hat{p}_f$	$\hat{p}_b$
0	0.0000	
1	0.1000	
2	0.2000	
3	0.3000	
4	0.4000	
5	0.5000	
6	0.6000	
7	0.7000	
8	0.8000	
9	0.9000	
10	1.0000	

Using properties of expectation, we can show that

$$E(\hat{p}_f) = p, \quad \text{Var}(\hat{p}_f) = \frac{p(1-p)}{10},$$

$$E(\hat{p}_b) = \frac{10p + \alpha}{10 + \alpha + \beta}, \quad \text{Var}(\hat{p}_b) = \frac{10p(1-p)}{(10 + \alpha + \beta)^2}$$

(11.8)

We see that $\text{Var}(\hat{p}_b)$ is smaller than $\text{Var}(\hat{p}_f)$. However, $\hat{p}_b$ is a biased estimator of p, whereas $\hat{p}_f$ is unbiased. It follows from Equations (11.6) and (11.8) that

$$\text{MSE}(\hat{p}_f) = \frac{p(1-p)}{10}$$

$$\text{MSE}(\hat{p}_b) = \frac{10p(1-p)}{(10 + \alpha + \beta)^2} + \left(\frac{10p + \alpha}{10 + \alpha + \beta} - p\right)^2$$

(11.9)

$$= \frac{[\alpha(1-p) - \beta p]^2 + 10p(1-p)}{(10 + \alpha + \beta)^2}$$

■ ■ ■ ■ ■ PARTING GLANCES

The Bayesian estimator $\hat{p}_b$ is based on both the observed data and the prior information. One interpretation of Equation (11.5) is that the prior information adds α base hits in $\alpha + \beta$ additional at bats to the observed data and then computes the traditional batting average. The effect of the prior can dramatically change the estimate of p for small n. You can probably see this in Table 11.1. However, if n is large relative to $\alpha + \beta$, the effect of the prior on the estimate is quite small.

■ ■ ■ ■ ■ REFERENCE

Official Baseball Rules (1998). St. Louis, MO: Sporting News Publishing.

■ ■ ■ ■ ■ ASSIGNMENT: CHOOSING AN ESTIMATOR

Include a plot of your prior density function with your responses.

1. What are your choices of α and β? What features of the plot of the prior density function made you think that these were good choices?
2. Plot MSE $(\hat{p}_f)$ and MSE $(\hat{p}_b)$ versus p for $0 \leq p \leq 1$ on the same plot. Does one estimator have a smaller mean squared error for all values of p? If not, determine the regions where each estimator is better.
3. Based on your comparison of the mean squared error functions and the information that Rocky Chew gave you, do you recommend using $\hat{p}_f$ or $\hat{p}_b$? Explain your choice.
4. If John D. Spurrier gets three hits in ten at bats, what is your estimate of p?

■ ■ ■ ■ ■ ASSIGNMENT: LOOKING AT THE MATHEMATICS

1. Use the results of Equation (11.7) and the laws of expectation to derive Equations (11.8) and (11.9).
2. Show that the Bayesian estimator $\hat{p}_b$ is a weighted average of $\hat{p}_f$ and $\alpha/(\alpha + \beta)$.
3. If X has a binomial distribution with parameters n and p, show that $\hat{p}_f$ is the maximum likelihood estimator of p.
4. If X has a binomial distribution with parameters n and p, show that $\hat{p}_f$ is the method of moments estimator of p.
5. If X has a binomial distribution with parameters n and p, show that $\hat{p}_f$ is the minimum variance unbiased estimator of p.

■ ■ ■ ■ ■ ASSIGNMENT: THE WRITTEN PRESENTATION

In your role as a statistician for On The Ball Consulting, write a formal report summarizing your work to be read by major-league scout Rocky Chew, his boss, and your boss. Your report should include the following:

1. A statement of the problem including the probability to be estimated
2. A description of the assumptions leading to the binomial distribution
3. A description of the prior information given to you by Rocky Chew
4. A description of the prior distribution that you chose for p and the rationale for its use
5. A description of the two estimation methods and example calculations
6. A comparison of the two estimation methods
7. A clear recommendation regarding which estimator to use
8. A technical appendix deriving the mean squared error formulas for the two estimators

For guidance, refer to Chapter 12: Strategies for Effective Written Reports.

■ ■ ■ ■ ■ ASSIGNMENT: THE ORAL PRESENTATION

In your role as a statistician for On The Ball Consulting, prepare a 5-minute oral presentation including professional-quality visual displays to present the results of your investigation to major-league scout Rocky Chew, his boss, and your boss. Your presentation should address items 1–7 of the Written Presentation assignment. You should deemphasize the mathematical development in your presentation. For guidance, refer to Chapter 13: Strategies for Effective Oral Presentations.

P A R T
II
Sharpening Nonstatistical Skills

Practicing statisticians don't generally design experiments or analyze data merely for their own benefit. They do this work to assist other parts of their employer's organization, whether it be top management, engineering, production, maintenance, the motor pool, etc. As a practicing statistician, you will seldom have sole authority to make decisions affecting these parts of the organization. To have an impact on the organization, you must master the nonstatistical skills of successfully communicating the implications of your statistical work and working effectively with nonstatisticians.

You will communicate through written and oral reports. The more effective you are in making these presentations, the greater your impact will be. Remember, the most ingenious experimental design and the finest data analysis have little value to the organization until you effectively present the results to others. Being able to write and speak effectively are important keys to your success as a statistician. You will be judged on your ability to communicate the results of your work.

Chapters 12–14 provide guidance on effective communications for the statistician. Written communication is discussed in Chapter 12, and oral communication is discussed in Chapter 13. Chapter 14 provides a brief introduction to PowerPoint®—a software product used for producing excellent visual aids. The emphasis in these chapters is on presenting applications of statistics to other fields rather than presenting theoretical developments in the field of statistics.

Chapter 15 provides guidance on working effectively with nonstatisticians. This is commonly called statistical consulting. In addition to reviewing different ways in which consulting is done, the chapter also discusses ethical considerations facing the practicing statistician.

To make an impact on an organization, you must first be hired. Chapter 16 is a primer on finding employment as a practicing statistician.

C H A P T E R
12

Strategies for Effective
Written Reports

■ ■ ■ ■ ■ **INTRODUCTION**

Being able to write effective reports is an important skill for statisticians. It takes considerable effort to produce professional written reports. Readers will judge you on what you write—not on what you think you wrote or what you know. As with most skills, written presentation skills are developed by learning the fundamentals, practicing, taking constructive criticism, and studying written reports. This chapter discusses some fundamentals of writing statistical reports. Additional sources are given at the end of this chapter. You may find some minor differences between the recommendations in this chapter and in other sources. You may wish to discuss these differences with your instructor.

■ ■ ■ ■ ■ **THE GOAL OF THE REPORT**

The statistician's goal is to *clearly inform the readers* of the results and implications of an experiment or of some statistical research. The statistician may also advocate that some action be taken based on these results. Above all, we wish to objectively inform the readers—not to entertain them, appeal to their emotions, or keep them in suspense. Our emphasis in technical writing is on logical development, accuracy, and ease of understanding. We should give the readers all the information they need to know about our work. There should be no room for differences in understanding among readers. Unlike an oral presentation, it is not always convenient for the readers to ask questions after reading the written report. Readers tend to be busy, so reports should be as concise as possible.

■ ■ ■ ■ ■ UNDERSTANDING THE INTENDED READERS

A written report has little value if the intended readers do not understand it. The writer must have some general information about those who are likely to read the report.

What are the backgrounds of the readers? Am I writing for accountants, engineers, operators, or statisticians? What types of education and work experience do they have? The choice of language and the points of emphasis may vary depending on the backgrounds of the readers.

What are the readers likely to know about the topic? If the readers are familiar with the data collection setting, it may be necessary to mention it only briefly. If not, you may need to provide considerable detail about the setting before the readers can understand the results of the experiment. For example, "We adjusted the pH in tank 5 from 4.5 to 4.0" may be completely understandable to those familiar with an industrial process but mean little to those unfamiliar with the process or the pH scale. There are similar issues regarding the readers' understanding of statistical methods and jargon. "The standard deviation of the part diameters is 0.32 microns" is clear to statisticians but means little to those unfamiliar with a standard deviation.

Structure your report based on your understanding of the intended readers' backgrounds and knowledge. If there are many potential readers, it may help to think of specific people who are representative of that group.

■ ■ ■ ■ ■ THE MAJOR SECTIONS OF THE REPORT

Written reports generally have several major sections. Section names vary in different settings. It is important to familiarize yourself with the customary written report style of your organization. The standard report sections we will consider include the following:

Title
Abstract or Executive Summary
Key Words and Phrases (optional)
Introduction
Materials and Methods
Results
Discussion
Conclusion

Acknowledgments (optional)
References
Appendixes (optional)
Glossary (optional)

Optional sections are included only if necessary to provide information or if required by your organization. In reports describing the development of new techniques in the field of statistics, one or more sections describing the research should replace the materials and methods section and the results section. Let's consider the purpose of each major section.

Title: Readers use the title to decide if they have interest in the report. The title should concisely, but completely, describe the subject of the report. It should be one or two lines. For example, a report of an experiment to evaluate the importance of various sources of variation in camshaft diameter measurements at the Green River factory might have the following title:

Impact of Operators, Machines, Cutting Oil Temperature, and
Measurement Error on Camshaft Diameter Measurements

The author's name and address and the date of the report are placed below the title. In some organizations, the author's telephone number and e-mail address are also given.

Abstract or Executive Summary: The abstract or executive summary is a self-contained summary of the objectives, results, and implications of the experiment. The term *abstract* is more common in scientific papers, and an *executive summary* is more common in business and industry. This report section allows the reader to quickly understand the major points of the report and to decide if the entire report needs to be read. Usually, a well-written abstract or executive summary contains 100 to 200 words, with no unnecessary language. The following is an abstract for the Green River camshaft experiment report:

The Green River facility has experienced continual problems with excess variability in the camshaft diameter measurements. Excess variability causes increased scrap, rework, and warranty claim costs. A quality improvement team identified the operator, the machine, cutting oil temperature, and measurement error as the most likely sources of variation. We conducted a three-factor experiment in May 1999 to evaluate these potential variation sources. Three operators and three machines were randomly selected. Controllers maintained the cutting oil temperature at the selected levels of 150, 180, and 210 degrees Fahrenheit. By carefully controlling cutting oil temperature at 150 degrees, we reduced the observed diameter variations by between 86% and 94% compared to historical data. Machine-to-machine differences, operator-to-operator differences, and measurement errors appear to be negligible. We recommend that oil temperature controllers be installed on all tooling machines at the Green River facility at a one-time cost of $20,000, with projected annual savings of $100,000 from decreased scrap, rework, and warranty costs.

This abstract briefly summarizes the major points of the report. A busy executive can get the basics without reading the entire report. An interested engineer may choose to read this report before reading ten others based on the information in the abstract. Other readers may learn that the study is not germane to their interests without reading the entire report.

Key Words and Phrases: This section is a list of key words and phrases in the report that are not in the title. These key words and phrases are used to index the report by its subject matter.

Introduction: The introduction gives the reader the necessary information to understand why the research was done and how the research fits into the existing knowledge base for the process or phenomenon being studied. This is often done, in part, by referencing other reports. It is important for the reader to understand how your research is different from the others. Don't reference reports that are only tangential to your work, but do reference highly related reports. The last paragraph of the introduction usually outlines the contents of the remaining sections.

The introduction for the Green River camshaft experiment report would include a description of the amount of variation historically observed in camshaft diameter measurements, a description of the impact of this variation on the company, a summary of the manufacturing process, a discussion of possible sources of variation, and references to other reports, if appropriate.

Materials and Methods: The materials and methods section, also known as the description of the experiment, gives a precise description of the experiment and the data collection. This description gives operational definitions for all variables; lists all major equipment used in the experiment, including that used in data measurement; and describes any randomization used in the experiment. The description must be precise enough that readers can reproduce your experiment. Some authors include the names of the statistical analyses in this section. No results should be placed in this section.

Results: In the results section, the author leads the reader through the statistical analyses. This section generally contains prose and tables or figures. The prose should fully describe the statistical methodology and guide the reader to the results, which can be in tables, figures, or prose. (Guidelines for tables and figures appear later in this chapter.) It is appropriate in this section to comment on the data analysis. However, discussion of the implications of the analysis comes in the discussion section.

Discussion: In the discussion section, the author discusses the implications of the statistical analyses and, if appropriate, makes recommendations based on these implications. The discussion should include as little statistical jargon as possible. It is also relevant to comment on any shortcomings or limitations of the present experiment and to suggest further experimentation in this section. It is important to clearly distinguish fact from conjecture. Do not overstate what the data tell you.

Conclusion: The conclusion reinforces the main points of the report. The conclusion differs from the abstract or executive summary in that the conclusion is read after, not before, reading the rest of the report. Thus, readers have a deeper understanding when reading the conclusion than when reading the abstract or executive summary.

Acknowledgments: The acknowledgments section acknowledges anyone who helped in the project or who provided funding for the project.

References: The references section (also known as the bibliography) lists complete citations for all books, articles, and reports referenced in the report. Different organizations have different styles for the references section. If the organization for whom you are writing does not require a particular style, follow the style used in the *Journal of the American Statistical Association*.

Appendixes: Material that will only be of interest to some readers or that will detract from the general readability of the report is placed in an appendix. Derivation of technical results and complete listings of data sets are often placed in an appendix.

Glossary: The glossary is an alphabetical list of definitions of specialized terms used in the report.

■ ■ ■ ■ ■ STYLE

Our goal is to inform the reader—a very busy person. He or she will read the report solely to gain the information, and probably needs to do this quickly. Writing style affects the effort required to gain the information. Technical writing must be precise, readable, and concise.

Precise writing has only one possible interpretation. Compare the vague statement "A *t*-test was done" with the precise statement "A one-sample *t*-test was done at the 5% significance level in an attempt to show that the experimental mean was larger than the control." The first statement is subject to many interpretations, whereas the second is not. Look for vague statements in your draft manuscript. Often, a minor change in wording will improve precision. The concrete noun "carpet tack" is more precise than the abstract noun "product."

Writing that is easy to read can be quickly understood. Technical writing experts give several suggestions for improving readability. Let's consider three suggestions. First, use understandable language. This includes defining any unfamiliar terms and jargon. Understanding comes slowly if the reader must constantly refer to a statistics book or dictionary.

Second, make the language as simple as possible without losing precision. For example, use the words *use* rather than *utilize, help* rather than *facilitate,* and *slowed* rather than *decelerated.* The goal is to inform, not to dazzle, the reader.

Third, use active rather than passive verbs and avoid prepositional phrases. Passive verbs are a form of "to be" followed by a verb ending in "ed." Prepositional phrases are prepositions followed by a noun or pronoun. The phrase "The length of the bolt was measured by the operator" is longer and not as easy to understand as "The operator measured the bolt's length."

Concise writing provides the necessary information in the fewest possible words. Most writing can be condensed without losing information. Several techniques make writing more concise. First, eliminate all words and phrases that add no information. For example, use "to" rather than "in order to," and "and so on" rather than "and so on and so forth." A good technical writing text will give more examples. Second, you don't have to include everything you know. Include the important and leave out or deemphasize the unimportant. Third, eliminate redundancies. For example, one graphical summary of a data set is probably sufficient. It is not necessary to include a dot plot, a histogram, a box plot, and a stem-and-leaf diagram. Finally, making your report more readable usually makes it more concise.

▪ ▪ ▪ ▪ ▪ TABLES AND FIGURES

Tables and figures are excellent devices for summarizing information. *Table* generally refers to columns of numbers, and *figure* refers to graphical displays, photographs, and diagrams. Structure tables and figures to tell their part of the story without reference to the prose. This is accomplished by carefully choosing the title, the labeling, and the amount of detail to include.

All tables and figures should have completely descriptive titles. The title "Box Plot" gives the reader no additional information. The title "Box Plot of Length Measurements of 100 Nominal 0.5-Inch Carpet Tacks to the Nearest 0.001 Inch" is much more descriptive.

Tables and figures should be no more complex than necessary to present the information. Extraneous clutter causes confusion. When summarizing data in a table or figure, specify the unit of measurement and clearly label all axes. Also, clearly indicate if an axis starts at something other than zero. This is often done by starting the axis at zero and indicating an axis break.

Number tables and figures sequentially for easy reference. The first table is Table 1 and the first figure is Figure 1. In long reports, include the section number in the table or figure number; for example, Table 3.1 is the first table in section 3.

While tables and figures should tell their part of the story without reference to the prose, refer to each figure and table by its number in the prose discussion. The reference should describe the information contained in the table or figure. Example references are "Figure 1 depicts a camshaft produced at the Green River facility" and "Table 1 provides an analysis of variance table for the camshaft diameter data." Place the table or figure as close as possible after the paragraph in which it is first referenced.

Often, you must choose between using a table or a figure in presenting statistical results. Tables are better for presenting fine detail, while figures are better for making comparisons or presenting trends.

■ ■ ■ ■ ■ WRITING THE FIRST DRAFT

After completing the experiment and analyzing the data, you have come to some conclusions. This is not the end of the road. You must now communicate these conclusions to others by writing your report.

For most statisticians, writing a professional report takes a considerable amount of time. Don't expect to produce a polished report in a single session or a single draft. It is best if you are well-rested and have large blocks of time to write. You will not always be that lucky. I prefer to write on a word processor, but some statisticians do early drafts with pencil and paper. In either case, double spacing the early draft aids in rewriting.

The writing process begins by organizing your thoughts logically in line with the goal of the report and the readers' backgrounds. Most people begin with an outline. The outline's major headings are the report's major section titles. The subheadings are the main points to be covered in each section. Writing the subheadings as complete sentences, rather than phrases, leads to better organization. After writing all subheadings, rearrange them into the most logical order. You may also have sub-subheadings. The outline should include a description of all tables and figures.

Once you have an outline, start writing the section that is the easiest to write. This is probably not the abstract or the introduction. Try to finish an entire section's draft in a single sitting. Don't worry about perfection; get something on paper. Revising is much easier than writing first drafts. Don't get bogged down in detail. If you have difficulty finding the proper wording, write "***" and go on. This will remind you that more work is required here. Word processors can quickly find these locations. Don't be too worried about writing style at this point.

When you finish the draft of the easiest section, go on to another easy section. Continue this process until you have a draft of the entire report. This will probably take several sittings. When you finish the last section, the hardest part is over.

■ ■ ■ ■ ■ REWRITING THE REPORT

Most of the effort in producing the first draft is to get something on paper. Careful rewriting is required to produce a professional report. This rewriting involves several steps and several sessions. If possible, let the first draft rest overnight before you begin rewriting. This will give you a clear head and will allow you to focus on what you wrote rather than on what you thought you wrote.

The purpose of the first rewriting session is to correct flaws in logic and fill in the missing material noted by "***." From the readers' viewpoint, ask the following questions: Is the presentation logical and consistent? Are there sentences that don't make sense? Are the tables and figures easily understood and professional-looking? Do all of the referenced sources appear in the reference section?

The other rewriting sessions deal with improving style. In these sessions, ensure that the writing is precise, readable, and concise. Look for grammatical errors, undefined or vague terms, jargon, needless complex language, passive verbs, prepositional phrases, words or figures that add no information, and redundancies. You may want to make a separate pass through the report looking for each type of flaw.

After making these revisions, check the manuscript with a spell checker and a grammar checker. You should also number the pages. At this point, you may think you are done, but you aren't. Have a trusted colleague critique the report. He or she may point out rough spots and find typos that eluded the spell checker.

Your final step is to slowly—very slowly—read the report one more time. Do this after not having looked at the report for at least one day. After all this rewriting, you may want to rush through the report. Force yourself to carefully read each word. You usually find that more improvements can be made.

As you see, writing statistical reports is hard work, but it is important to do it well. You are responsible for what you write, and you will be judged by it.

■ ■ ■ ■ ■ IMPROVING AS A WRITER

Throughout your career, you should continually improve your writing skills. As with most skills, there are three ways to improve: practice, observe, and read.

Practice: Most professional statisticians get plenty of practice. You will write numerous reports. Strive for continual improvement in your writing and take advantage of opportunities for additional training in technical writing.

Observe other reports: You will read numerous reports. Learn from each of these. What aspects of a report are particularly good and what aspects detract from the report? Consider incorporating the good into your style and avoid the bad.

Read about technical writing: Most libraries and bookstores have several technical writing books. Alley (1996) is a good general-purpose text on scientific writing. You should also have a good style manual in your personal library. *The Chicago Manual of Style* (1993) is one of several excellent style manuals.

▪ ▪ ▪ ▪ ▪ REFERENCES

Alley, Michael (1996), *The Craft of Scientific Writing* (3rd ed.). New York: Springer-Verlag.

The Chicago Manual of Style (14th ed.) (1993). Chicago: University of Chicago Press.

▪ ▪ ▪ ▪ ▪ WRITTEN REPORT CHECKLIST

The following checklist gives reminders for preparing and assessing your written reports.

Report Goal
___ Do you have a clearly defined goal or goals?

Intended Readers
___ What are the backgrounds of the intended readers?
___ What are they likely to know about the topic?

Outline
___ Have you developed an outline?
___ Are the major sections listed as major outline headings?
___ Does each major section have complete sentence subheadings?

Title
___ Does the title completely describe the subject of the report?
___ Are there any unnecessary words?
___ Are your name, address, and the date included under the title?

Abstract or Executive Summary
___ Is the abstract or executive summary understandable without reading the rest of the report?
___ Does it summarize the objectives, results, and implications of the research?
___ Are there any unnecessary words?

Key Words and Phrases (optional)
___ Are all key words and phrases that are not in the title included in the list?

___ Are the key words and phrases in alphabetical order, separated by commas?

Introduction
___ Can the reader understand why the research was done?
___ Can the reader understand how this research fits into existing knowledge?
___ Is there appropriate referencing to related reports and/or research?
___ Can the reader understand how this research differs from related research?
___ Does the final paragraph outline the remainder of the report?

Materials and Methods
___ Are the experimental methods clearly described?
___ Are all operational definitions clearly stated?
___ Are all major pieces of equipment listed?
___ Are the data collection methods clearly described?
___ Is all randomization described?
___ Would the reader be able to reproduce your experiment?

Results
___ Are the statistical methods and results clearly presented?
___ Are all tables and figures clearly introduced in the prose?
___ Do the tables and figures tell their part of the story without reference to the prose?
___ Are the tables and figures neat and free of clutter?
___ Are all tables and figures numbered?
___ Do all tables and figures have descriptive titles?
___ Are all axes clearly labeled?
___ Are axes that don't start at zero clearly noted?
___ Are units of measurement clearly defined?

Discussion
___ Are the implications of the statistical analyses clearly discussed?
___ Have you made all appropriate recommendations based on the results?
___ Have you pointed out any shortcomings or limitations of the present research?
___ Is your discussion free of statistical jargon?
___ Have you clearly distinguished conjecture from fact?
___ Is your discussion consistent with the statistical results?

Conclusion
___ Have you summarized the main points of your report?

Acknowledgments (optional)
___ Have you acknowledged all individuals, other than co-authors, who contributed to the experiment or report?

References
___ Have you given complete citations for all sources referenced in your report?
___ Are the citations consistent with the style you are using?

Appendixes (optional)
___ Are the appendixes appropriately titled and identified by number or letter?
___ Is each appendix referred to in the body of the report?

Glossary (optional)
___ Are all specialized terms defined?

Style
___ Does any of your writing have more than one possible interpretation?
___ Have you used any terms or jargon that the readers will not understand?
___ Have you used any big words when smaller words would carry the same meaning? (Use the word *use* rather than *utilize*.)
___ Have you avoided unnecessary passive verbs and prepositional phrases? (Forms of *to be* followed by a word ending in *ed;* of, for, and by)
___ Have you eliminated words and phrases that add no more meaning? (Use *to* rather than *in order to*.)
___ Have you eliminated all unimportant information?
___ Have you eliminated all redundant information and figures?

Final Checks
___ Are the pages numbered?
___ Have you used a spell checker?
___ Has a colleague critiqued your report?
___ Have you carefully proofread your report after making all changes?

C H A P T E R
13
Strategies for Effective Oral Presentations

■ ■ ■ ■ ■ **INTRODUCTION**

Being able to make effective oral presentations is an important skill for statisticians. The thought of making an oral presentation may make you somewhat uneasy. This is normal. You cannot expect to be an expert speaker in your first attempts any more than a beginning typist or golfer can expect initial professional performances. As with most skills, you can develop your oral presentation skills by learning the fundamentals, practicing, taking constructive criticism, and watching others. This chapter discusses some oral presentation fundamentals. You can find references to additional information at the end of this chapter.

■ ■ ■ ■ ■ **THE PRESENTATION GOAL**

The first step in preparing an oral presentation is to focus on the presentation goal or goals. A goal of a statistician's presentation is usually to inform the audience of the results and implications of a data collection activity such as a survey or an experiment. Another goal may be to advocate that some action be taken based on these results. This action could be a change in a production process, a change in a marketing plan, additional experimentation, and so on. The former goal is to create understanding, while the latter is to change behavior.

■ ■ ■ ■ ■ UNDERSTANDING THE AUDIENCE

Communication involves both the speaker and the audience. An elegant speech has no value if the audience does not understand it. The second step in preparing an oral presentation is learning about the expected audience. The person asking you to speak is usually a good source of information. There are several questions about the audience that need to be answered. The answers to these questions will help you design your presentation to achieve the maximum possible impact with that audience.

What are the backgrounds of the audience? Am I speaking to accountants, engineers, operators, or statisticians? The choice of language and the points of emphasis may vary depending on the background of the audience.

What are they likely to know about the topic? If the audience is very familiar with the setting in which the data were collected, it may be necessary to mention it only briefly. If not, you may need to speak in depth about the setting before the audience can understand the experimental results. For example, "We adjusted the pH in tank 5 from 4.5 to 4.0" may be completely understandable to those familiar with an industrial process but mean little to those unfamiliar with the process or the pH scale. There are similar issues regarding the understanding of statistical methods and jargon. "The standard deviation of part diameters was 0.32 microns" is clear to statisticians but means nothing to those unfamiliar with a standard deviation.

What do they know about you? The effective speaker must be credible. For the statistician, credibility comes from educational background and work experience. If you do not have a close working relationship with the audience, it will be necessary for you or the person introducing you to provide some background information about you.

How many people are expected? The size of the audience may dictate whether a microphone is needed, when questions will be allowed, and the style of visual aids and handouts that can be used. For example, flip charts and handheld props might be effective visual aids for a small group but ineffective for large audiences.

■ ■ ■ ■ ■ UNDERSTANDING THE GROUND RULES

The third step in preparing an oral presentation is clarifying logistical and procedural details.

How much time has been allocated for your presentation? This amount of time is generally the total time allocated for someone to introduce you, for your presentation, and for a question-and-answer period. In this case, use about 80% of the allocated time for your presentation. In situations where numerous questions are anticipated or discussion is to be encouraged, you would use less time for the presentation. It is bad form to use more than the allocated time. In some settings, you will be forced to stop at the end of the allocated time even if you are not finished.

What are the facilities? Learn the approximate size and special features of the room in advance. Also, find out what equipment will be available. It can be disastrous to plan a presentation assuming certain equipment will be available and then discover at the last minute that it is not available or cannot be used effectively in the room. Visit the room and try out the equipment as far in advance as possible. Learn how to use the projector before, not during, your presentation!

When will questions be asked? It is easier to keep your presentation on schedule if questions are asked at the end. In some settings, there is a tradition of asking questions during the presentation. If this is done, give short answers unless there are no time constraints. In a business setting, it is unwise to make your superiors wait to have their questions answered.

Will other speakers speak on related topics before me? If so, clarify with them what will have been said before you speak. Otherwise, you may discover at the last minute that your entire presentation has already been given.

Is a written version of my presentation needed? Written versions contain more details than the oral presentation and allow the readers to read your work at their own pace. Written documentation is also valuable as a point of reference in future work. If a written version of the presentation is needed, follow the suggestions in Chapter 12. Some organizations require speakers to submit a written version of their presentation.

Is a title needed? If your presentation will be publicized through flyers, a program, or an agenda, you will need a presentation title. You also need a title if the presentation will be referenced by authors. Otherwise, a title is optional.

■ ■ ■ ■ ■ THE MAJOR SECTIONS OF THE PRESENTATION

The next step in preparing the oral presentation is developing a presentation plan. Begin the development by listing the major sections of your presentation. The following major sections will be appropriate for most presentations of data collection activities by a statistician:

Greeting and Introduction
Preview
Background
Description of the Data Collection Activity
Results of the Data Analysis
Conclusions and Recommendations

Let's consider the purpose of each major section.

Greeting and Introduction: In the greeting and introduction, you get the audience's attention, establish a spirit of goodwill, establish your credibility, and provide a base for the rest of your presentation. The following greeting and introduction would be appropriate for a statistician reporting on an industrial experiment to senior management.

> Good morning! I am Willis Ward, a statistician at our Green River facility. The Green River mission is to produce high-quality camshafts. Our facility has experienced continual problems with excess variability in the diameters of these shafts. This variation is costing our company over $100,000 annually. Mechanical engineer Linda Wright, machinist J. W. Johnson, and I have studied this problem. My presentation goals are to describe our study and to recommend an action to substantially reduce this variability.

Notice how the speaker accomplishes the greeting, the establishment of credibility, and the statement of goals in a relatively small number of words. Efficient use of language is important to conserve time for other parts of the presentation and to maintain the audience's attention.

To help establish goodwill and credibility, smile and make solid eye contact with the audience during the greeting and introduction. The focus should be strictly on you, with no distractions from visual aids or handouts.

Preview: In the preview, you give your order of presentation. This allows the audience to focus on each section as it is presented. During the preview, establish verbal signposts. These are key phrases describing each section. The following preview would be appropriate for Willis Ward's presentation:

> I will begin with some **background information** about camshaft production. I will then describe our **experiment** and the **results**. Then, I'll give our **conclusions and recommendations**. Finally, I will receive your **questions and comments**.

The words in boldface are the verbal signposts. Each section's verbal signpost should be repeated as you make the transition to that part of your presentation. For example, Willis Ward should begin the next section of his presentation by saying, "Now let me give some **background information**."

Verbal signposts can be reinforced as visual signposts through the effective use of visual aids. Figure 13.1 depicts a visual aid that reinforces the verbal signposts.

Background: In the background, you give the audience the information they need to understand the data collection activity and its relevance. In developing the background, keep in mind what the audience knows and does not know about the project. Don't bore the audience with facts they already know, nor lose them because they lack necessary information.

> *Preview*
>
> 1. Background
>
> 2. The Experiment
>
> 3. The Results
>
> 4. Conclusions and Recommendations
>
> 5. Questions and Comments

FIGURE 13.1   A visual aid that reinforces verbal signposts

Figure 13.2 displays an outline for the background in Willis Ward's presentation. In the background, Willis Ward will describe the problem, explain why it is important, and give the project team's rationale leading up to the experiment. Each step in this outline can be enhanced with visual aids.

Most information in the background is nonstatistical. You can learn about the nonstatistical aspects of the project from your colleagues or other sources.

> A. A camshaft (real, model, or picture) made at Green River is shown.
>
> B. The process of measuring shaft diameter is explained or demonstrated.
>
> C. The amount of variation present in camshaft diameters under historical operating conditions is explained with graphical and numerical summaries.
>
> D. The cost of the variation in terms of scrap, rework, and warranty costs is quantified.
>
> E. The camshaft manufacturing process is described.
>
> F. The possible sources of variation in the diameter measurements are reviewed.
>
> G. The project team's rationale for studying specific possible variation sources is presented.

FIGURE 13.2   Willis Ward's background outline

Description of the Data Collection Activity: In the description of the data collection activity, you specify how the data were collected. The following is the description given by Willis Ward:

Let me now describe our **experiment**. It was conducted at the Green River facility during the May 26, 1999, day shift. The experiment involved three variables: machinists, tooling machines, and cutting oil temperature. There were three randomly selected machinists, three randomly selected machines, and three levels of oil temperature: 150, 180, and 210 degrees Fahrenheit. The uncontrolled oil temperature has been known to vary from 150 to 210 degrees. The oil temperature was controlled using a DWW Model 345 controller. Using blanks from the same lot, each machinist produced five camshafts on each machine for each oil temperature. Thus, each machinist produced $3 \times 3 \times 5 = 45$ camshafts. The diameter of each camshaft was measured twice.

Results of the Data Analysis: In this section, you summarize the data analysis using language that is understandable to the audience. It serves no purpose to blow the audience away with a statistical presentation that they don't have the background to understand. If you feel a strong need to present complicated details of an analysis, put them in the written version of the presentation or in a handout, but don't dwell on them in the oral presentation. You should, however, be ready to talk about them if you are asked to do so. Willis Ward might give the following results of the data analysis for his experiment to senior management:

I will now summarize the **results** of the data analysis. The data suggest there is little variation due to measurement error, to machine-to-machine differences, or to machinist-to-machinist differences. The data suggest that oil temperature is a major source of the variation in the camshaft diameters. The visual aid shows histograms of shaft diameter data from historical data and at each of the three oil temperatures. As you can see, the variation for each of the three controlled temperatures is considerably less than that for the historical data. The least variation occurs when the oil temperature is 150 degrees. This setting produces a reduction of between 86 and 94% in the diameter variation compared to historical data. Keeping the oil temperature at 150 degrees should decrease our scrap, rework, and warranty costs by $100,000 per year. Details of the data and cost analyses are included as appendixes of the written report.

Conclusions and Recommendations: In the conclusions and recommendations, you bring everything together. Specifically, you summarize what you have said, recommend specific actions, and invite questions from the audience. The following conclusions and recommendations would be appropriate for Willis Ward's presentation:

In **conclusion**, our project team performed an experiment to study the effect of machines, machinists, and cutting oil temperature on the variability of camshaft diameters produced at the Green River facility. We learned that the oil temperature has a major effect on variability. Keeping the temperature at 150 degrees Fahrenheit reduces our variation by about 90%. There is little machine-to-machine or machinist-to-machinist variation. We recommend an expenditure of $20,000 to equip all tooling machines at Green River with oil temperature controllers. We project an annual savings of $100,000 from decreased scrap, rework, and warranty claims. I thank you for this opportunity to meet with you. I will be happy to take your **questions and comments**.

The conclusions and recommendations are the punch lines of the presentation. You want the audience to remember them, even if they remember nothing else. It is crucial that you maintain a high energy level to the end. Compare Willis Ward's conclusions and recommendations to those of the weak finish:

Well, I guess that's about all I have to say. Uh, it must be time to quit. I hope, you know, I haven't bored you too much and that you got something from my talk.

The weak finish merely fills (wastes) time and provides no useful information.

▪ ▪ ▪ ▪ ▪ **THE PRESENTATION OUTLINE**

The development of the presentation plan continues with the creation of a presentation outline. The major levels of the outline are the presentation's six major sections. It is helpful to make a time allocation for each major section. This allocation, which can be adjusted later if necessary, is highly dependent on the nature of the data collection activity and the background of the audience. A tentative allocation for Willis Ward's 10-minute presentation to senior management is given in Figure 13.3.

The major levels will usually have sublevels corresponding to important points within that section. For example, Willis Ward has seven

---

I. Greeting and Introduction (0.5 minute)

II. Preview (0.5 minute)

III. Background (6 minutes)

IV. Description of the Data Collection Activity (1 minute)

V. Results of the Data Analysis (1 minute)

VI. Conclusions and Recommendations (1 minute)

FIGURE 13.3 Outline of the major levels with tentative time allocations

sublevels under the major level Background (see Figure 13.2). Some of these sublevels will have sub-sublevels. The steps of the camshaft manufacturing process are sub-sublevels of sublevel E (see Figure 13.2) of Willis Ward's major level Background.

Experts in public speaking recommend the use of complete sentences in the outline. This encourages more detailed thinking in the development of the outline.

As you develop your outline, you will likely struggle with the following questions:

What topics need to be covered?
How much time is available for each topic?

Your first responses to these questions may be that you have twice as much to say as you have time. If so, you may be tempted to talk twice as fast as normal. This ignores another important question

How much can the audience understand about the topic in the allotted time?

Remember, the goal is to inform and possibly advocate change. Showing the audience how fast you can talk is not a goal!

If you have more to say than time allows, you must either prune the outline or find more efficient ways to present ideas. For example, using a carefully planned visual aid may allow you to explain a production process much more quickly. Efficient use of language also allows you to accomplish more in a fixed amount of time. Efficient use of language is discussed next.

■ ■ ■ ■ ■ THE PRESENTATION STYLE

You must develop your own oral presentation style. You may have different styles for different presentations. In developing your style, remember that the focus in an oral presentation should be on you. You may wish to use visual aids, but these should be *aids* and not the focus of the presentation. There are a number of questions to consider in developing your style for a presentation.

What presentation mode should be used? Oral presentations can be read or given from memory from a prepared manuscript, given from notes (extemporaneous), or given without preparation (impromptu). Unless you are very experienced and highly skilled in giving public readings, reading a presentation tends to bore audiences. In a memorized presentation, the speaker often uses so much energy to make sure the words come out exactly right that he or she loses contact with the audience and appears unenthusiastic. Impromptu presentations tend to be highly unorganized, especially for an inexperienced speaker. Most public speaking experts suggest a well rehearsed speech given from notes. Speakers are more spontaneous and relaxed using this presentation mode. The notes can be key words or phrases taken from the presentation outline. Note cards should be numbered to keep them in order.

Does my appearance matter? Yes! Remember, you want the focus on you, and that focus should be positive. In most business settings, male speakers should wear a suit, and female speakers should wear a suit or business dress. Save your flamboyant clothes for other occasions.

Your clothes are only part of your appearance. Be in control of your body language. Strive for good posture, but don't be a statue. Balance your weight on both feet. Avoid the teeter-totter effect of constantly shifting your weight from one foot to the other. Also, avoid pacing. Have a purpose for any walking you do during your presentation.

Another important part of body control is the positioning of your arms. It is best to let them hang loosely at your sides, leaving them in a perfect position for gesturing. Alternative positioning of the arms, such as folded across your chest, wrapped behind your back, or wrapped around your neck, distract the audience.

A final part of body control deals with your face. Strive for an expressive face complete with a smile. Maintain eye contact with individuals throughout the audience. This improves your credibility and gives you instant feedback on their reaction to your presentation.

What are some key aspects of the oral delivery? First, be enthusiastic. You are selling yourself and the statistics profession through your presentation. If you can't get excited about your presentation, why should the audience? Enthusiasm leads to natural variation in pitch and helps you avoid speaking in a monotone.

Second, speak with enough volume to be easily heard at the back of the room. If you must strain to do this, you need a microphone. A friend at the back of the room can let you know if you have enough volume.

Third, clearly enunciate each word. Don't mumble!

Fourth, speak at a rate that can be comfortably understood. A friend can help you judge if you are speaking too rapidly or too slowly by listening to a rehearsal.

Fifth, remove clutter from your presentation. Clutter is words and phrases that explain nothing. Examples are the inclusion of "You know," "Like," "Uh," and "Okay" in sentences. Say "She said it" rather than "Okay, she, you know, like said it." Other examples of clutter are unnecessary extra words, such as "whether or not" in place of "whether" and "and so on and so forth" rather than "and so on."

It is helpful to compare bad and excellent speakers. What makes the bad speakers bad and the excellent speakers excellent? Perhaps you have heard a speech by Mr. Boring. He begins by apologizing for not being a good speaker. He then drones through the presentation, giving the clear impression that he would rather be somewhere else. He is difficult to understand because he mumbles and his volume drops off at the end of many sentences. Few people remember what he said.

Contrast Mr. Boring with excellent speakers you have heard. Excellent speakers are enthusiastic. The audience feels their excitement. It is easy to listen to them because you don't have to strain to hear and you don't have to work to decipher the words. Every syllable comes through loud and clear. It is a pleasure to hear them.

■ ■ ■ ■ ■ ■ VISUAL AIDS

Visual aids can be very helpful in showing a product, explaining a process, or reinforcing ideas. Visual aids include overhead transparencies, photographic slides, computer image projections, charts, and props. Overhead transparencies and computer image projections are the most common visual aids used by statisticians. As computer image projection allows for dynamic visual aids, this method is becoming increasingly popular.

The purpose of visual aids is to help the audience understand the speaker's message. They *aid,* but don't replace, the speaker. Using too many visual aids overloads the audience and removes the focus from the speaker. Don't put every idea on an overhead transparency. Use them for specific purposes.

To be effective, visual aids must be easily seen, legible, and easily understood. They can be enhanced with the use of color. These points may seem obvious, but many statisticians use visual aids that lack these properties. Let's investigate each property in more detail.

Seen by the entire audience: Small handheld props or a flip chart can work well in a small room but can't be seen from the back of a large room. When using a screen to display images, position it so that everyone can see, and don't stand where you block the view of anyone in the audience. It may be necessary for someone to assist you in displaying the visual aids.

Legible: Legibility involves both neatness and the size of writing and figures. In many business settings, it is expected that visual aids will be prepared using specialized presentation programs such as PowerPoint. These programs make it very easy to produce high-quality visual aids. In other settings, *neat* hand printed visual aids are quite acceptable. If your printing tends to be sloppy, use computer-generated visual aids.

With both computer-generated and hand printed visual aids, the characters should be large enough to be easily read from the back of the room. For computer-generated transparencies, use Arial or Helvetica font,

at least 24 point type,

and a laser printer. The size of 24 point type is a guide for hand printed characters on transparencies. Characters on flip charts need to be much larger. Again, the key is having characters large enough that they can easily be read from the back of the room.

When hand printing transparencies, place the transparency on a sheet of lined paper to assist you in writing in straight lines and maintaining uniform character size. Don't rush! It takes time to produce neat transparencies. Special pens are available for writing on transparencies. Use permanent pens when creating transparencies. Nonpermanent pens are handy for annotating transparencies during a presentation if the transparencies will be reused.

Easily understood: The secret to creating understandable visual aids is to keep them simple. There are several ways to simplify a visual aid. Limit each aid to one topic. You can use a series of aids to present a sequence of ideas. Use phrases rather than complete sentences. You will be saying the sentences. A key phrase will reinforce your words. Make your ideas stand out by leaving lots of white space and by using bullets to highlight items in a list. Figure 13.4 demonstrates these points in a visual aid on simplifying visual aids.

Simplifying Visual Aids

- One topic per visual aid

- Phrases, not complete sentences

- Leave white spaces

- Bullets highlight items

FIGURE 13.4   Methods of simplifying visual aids

The information in Figure 13.4 could be presented in a series of visual aids. Each visual aid in the series could highlight or magnify a particular discussion item. Figure 13.5 highlights the third item in the list.

Simplifying Visual Aids

- One topic per visual aid

- Phrases, not complete sentences

- **Leave white spaces**

- Bullets highlight items

FIGURE 13.5   Example visual aid that highlights one item

Complex figures and large tables of numbers are particularly difficult for audiences to understand. It is better to place these in a written report or handout. Simplify figures, as much as possible, without misleading the audience. It may not be necessary to display all the numbers from a table in the written report. Perhaps the major point from the table can be presented with a graph. Tables and figures are more easily understood if a descriptive title is included. Label graph axes with variable names and measurement units.

Use of color: Color can liven up any visual aid. Color is particularly helpful in graphical summaries of data and in diagrams of systems. Color can also be used for visual cues regarding the organization of your talk. For example, in presenting the visual aid in Figure 13.4 you might use one color for the title line and a different color for the four items in the list. Alternatively, you might use one color for section headings and a different color for ideas within that section. It is much easier to use color when projecting images directly from a computer.

Hand printed visual aids have the advantage that one can use different colors by changing pens. Making computer-generated, color transparencies requires access to at least a color printer and, perhaps, a color copier. If you can't make computer-generated, color transparencies, you do have the option of highlighting black-and-white transparencies by underlining and boxing key ideas in color.

Now that we understand the desirable properties for visual aids, let's consider some details of using them during your presentation.

Practice with the equipment: Pity the poor speaker who displays his first transparency only to discover that the audience is chuckling because the image is upside down, backward, and out of focus. The speaker then quickly tries to make the image right. This fumbling wastes precious presentation time and distracts the audience.

Don't let this happen to you, so arrive early, be sure you know how to use the equipment, and be sure that the image is in focus. On most overhead projectors, the top of the transparency is placed in the direction away from the screen.

Image alignment: Be sure the image is aligned properly on the screen. The entire image should be on the screen and the writing should be parallel to the top of the screen. Transparencies tend to move when touched. So touch them only when necessary, and realign the image if the transparency moves.

Keeping visual aids in order: In the excitement of preparing and making a presentation, it is easy to get the visual aids out of order. Not only is this embarrassing, but sorting visual aids distracts the audience and wastes presentation time.

Number your transparencies or photographic slides. This allows you to quickly check their order prior to the presentation. During the question-and-answer period, the audience can refer to a visual aid by its number and you can use the number to locate the visual aid. It is also convenient to annotate your presentation outline and notes with the numbers of the visual aids that will be used.

Some speakers put their transparencies in covers that fit in a three-ring binder. They remove the transparencies from the binder and return them to it immediately before and after their use. This is an excellent way to keep transparencies in order.

Proofread your visual aids: It is distracting to the speaker and the audience to discover an error in a visual aid during a presentation. Carefully proofread all visual aids well in advance of the presentation.

Don't forget about the audience: Look at and speak to the audience, not the visual aids. If you can't give your presentation without constantly looking at your visual aids, you need more rehearsal.

■ ■ ■ ■ ■ HANDOUTS

Unlike a written presentation which can be reread, the audience seldom gets to see an oral presentation more than once. Over time, the audience will forget much of what you present. Always have a handout to help the audience remember your vital points. Your name, address, e-mail address, and the date and location of the presentation and perhaps your phone number should be on the handout. Handouts take a variety of forms.

The "written manuscript" handout has the advantage of providing all details. It has the disadvantage that only highly motivated members of the audience will read it. It can be difficult to produce enough copies and transport them to the meeting site. For large audiences, it is best to say that a written manuscript is available on request.

The "copy of all visual aids" handout has the advantage of being easy to produce. It is only necessary to photocopy your transparencies. You will probably want to reduce the images to fit more than one on a page. You may also want to leave white space for note taking. This type of handout gives the framework of the presentation and provides reminders of what you said.

The "summary of main points and references" handout is particularly good if you have a few main points that you want the audience to remember or if you are giving a talk that reviews the work of several investigators in a field.

You need a plan for distributing the handouts. Do not attempt to distribute them during your presentation. Either have them distributed prior to your presentation or announce that handouts will be available after your presentation.

■ ■ ■ ■ ■ REHEARSING THE PRESENTATION

Another important part of oral presentation preparation is having a series of rehearsals. It is ideal to start the rehearsal process several days before the presentation date. As with a good stew, oral presentations need considerable "cooking" time. It takes time to get comfortable with your presentation and to smooth out any rough edges. If you don't have several days, do as much of the following as possible.

The rehearsal process begins by reading the presentation outline aloud several times. These readings help you get the presentation order clearly in mind. When you are comfortable with the presentation order, prepare your presentation notes. Unlike the complete sentence format of the outline, presentation notes consist of key words or phrases that remind you what comes next. The notes should be written in large print on note cards.

The next step in the rehearsal process is to practice your complete presentation using the notes. Also, use your visual aids, if they are ready. It is common for this rehearsal to be very rough. Part of the "cooking" process is discovering the portions of the presentation that require extra attention. Regardless of how rough the first attempt is, force yourself to complete the entire presentation. You will probably want to do this rehearsal in private.

After you complete this stage, think about the parts that were particularly rough. Perhaps, you had difficulty getting started, making transitions between major ideas, or explaining a particular idea. All of this is to be expected in this stage. As you consider the rough spots, think of ways to make these parts smoother. If you are having difficulty getting started, start with an enthusiastic greeting and have the first few sentences firmly in mind. If you are having difficulty with transitions, think of sentences such as "Now that we understand how the data were collected, let's move to the results." If you are having difficulty explaining an idea, think of alternative wording or perhaps add a helpful visual aid. After rethinking each rough part, rehearse that part a few times.

If you had several rough spots, repeat the rehearsal of the complete presentation in private. If there are just a few rough spots, you are ready to move to the next stage of rehearsal. In this stage, give your complete presentation to a trusted friend, using all visual aids and, if possible, record the presentation. Choose a friend who will give you honest, constructive criticism. You may not enjoy receiving criticism, but listen intently to your friend's comments. They will probably be helpful. When you help friends improve their presentations, remember to praise the good parts as well as suggest improvements.

You want to accomplish three things in this stage. First, get comments from your friend on parts of the presentation that were difficult to understand, your presentation style, and your visual aids. Second, measure the amounts of time used in the entire presentation and in each major part. This is best done by reviewing the recording but can also be done by your friend. Third, identify clutter in your presentation. Both your friend and the recording can be useful in identifying clutter.

It is now time to fine-tune the presentation to meet your time requirements. If your timed rehearsal exceeded the time limit, you have to cut. Don't make the mistake of deciding to talk faster! First, cut the clutter. This may dramatically shorten and improve your presentation. If your presentation is still too long after all of the clutter has been removed, you need to find ways to present some topics more efficiently or delete some of the least important material. An effective visual aid can often be helpful in presenting material more efficiently.

If the timed rehearsal was shorter than the time limit, ask your friend if any parts of the presentation needed additional explanation. If not, you have the luxury of adding additional material or allowing extra time for discussion.

After adjusting your presentation in response to your friend's comments and to meet time restrictions, you are ready for the final rehearsal. It is best to have this rehearsal in a room similar to the one where you will make the presentation. Your friend should sit at the back of the room to make sure you can be heard and that the visual aids can be seen. He or she should also time the presentation. Your friend should again give you feedback after this rehearsal.

■ ■ ■ ■ ■ **OVERCOMING FEAR**

It is natural to be somewhat uneasy before giving an oral presentation. Some beginning speakers are so uneasy that they have difficulty making their presentation. These speakers need strategies for overcoming fear. Sprague (1995) presents some excellent strategies. Let's review and expand on some of her ideas.

Analyze your fear as specifically as possible: You may say, "Giving a presentation scares me to death." This statement is too general to be helpful. It is important to list exactly what you are afraid of. Specifics can be dealt with, generalities cannot!

For example, you may be afraid that you will forget your speech. This fear can be overcome by more rehearsal and a good set of presentation notes. Most experienced speakers occasionally get lost but quickly get back on track by looking at their notes. A final review of your notes a few minutes before you speak can be very helpful.

You may be afraid that your knees will shake. It may be difficult to keep your knees from shaking but the audience doesn't have to see them. Podiums or loose-fitting clothes will hide shaking knees.

You may be afraid that you will put the transparencies on the projector upside down and backward. A little practice with the projection device will generally take away this fear. Moreover, if the image appears upside down and backwards, just smile; the world will not end.

You may be afraid that you will be asked a question that you cannot answer. It is perfectly appropriate to respond "I don't know" to a question. If it is practical, add "Let me find out, and get back to you."

Most specific fears can be overcome by thorough preparation. It is also comforting that most of these fears go away as you gain experience.

Replace negative thoughts with positive ones: Attitude can be a major factor in determining success or failure in most tasks. As you approach the task of giving an oral presentation, be positive. Rather than concentrating on all the things that might go wrong, realize that you are prepared and have important information.

Many golfers improve their performance by visualizing a perfect shot before they hit the ball. This technique, which is also helpful in preparation for an oral presentation, replaces all negative thoughts with positive ones. Visualize yourself giving an excellent oral presentation—so good that the audience triumphantly carries you around the room on their shoulders.

Use relaxation techniques: It is normal to become nervous in the minutes before your presentation. This nervousness may display itself through sweaty palms, increased heart rate, muscle tightness, and so on. This type of nervousness can be greatly reduced by taking some deep breaths and slowly exhaling. You can also visualize parts of your body becoming increasingly relaxed. These techniques get many women through childbirth without medication—you're only giving an oral presentation! If it is appropriate, you may want to take a short walk or do some light physical exercise to burn off some nervous energy.

Don't quit: The beginning of a presentation is stressful to most speakers. The speaker generally gains confidence and becomes more relaxed as the presentation proceeds. Also, the speaker gains confidence as he or she gains experience in giving presentations. It may be necessary for you to "tough it out" through a few uneasy moments to develop the confidence you will need in the future.

■ ■ ■ ■ ■ THE QUESTION-AND-ANSWER PERIOD

Most presentations end with a question-and-answer period. This useful period allows the audience to seek clarification and provide input. It is often difficult to predict the nature of the questions and comments. You must rely on your knowledge of the topic and be willing to say "I don't know." Not knowing what is coming can be unsettling, but have a positive attitude. It is your opportunity to learn from the audience. The following techniques make the question-and-answer period more effective.

Understand the question before answering: Listen carefully to the question. A confused or nervous questioner may ask a poorly worded question. State your understanding of the question before giving your answer. This gives the questioner an opportunity to correct you if you misunderstood the question. It also lets the entire audience hear the question.

Collect your thoughts before giving an answer: Pause briefly before answering. Use the time to organize your answer. This technique reduces rambling.

Give your answer to the entire audience: Although the question comes from a particular person, give your answer to the entire audience. Many of them may have had the same question. When giving your answer, make eye contact with different parts of the audience. After completing your answer, make eye contact with the questioner and say, "I hope that answers your question."

Dealing with ego: Occasionally, a member of the audience will make a comment for the sole purpose of trying to appear important. If the comment is not appropriate, thank him or her for the input and perhaps offer to talk more after the presentation.

Be approachable: Show the audience that you are interested in their questions. Maintain eye contact with the audience, smile, and be polite to all questioners.

■ ■ ■ ■ ■ IMPROVING AS A SPEAKER

You should continually improve your presentation skills throughout your career. As with most skills, there are three ways to improve: practice, observe, and read.

Practice: Give oral presentations regularly. There are several opportunities for practice. Most major cities have a chapter of Toastmasters—a club to help members improve their public speaking skills. Alternatively, most civic and religious organizations are constantly looking for officers. These positions frequently allow you to make oral presentations.

Observe other speakers: You have numerous opportunities to observe oral presentations. Learn from each of these. What aspects of the presentation are particularly good? What aspects detract from the presentation? Consider incorporating the good into your style and avoid the bad. Pay special attention to the methods speakers use to make technical material easy to understand.

Read about public speaking: There are numerous works on oral presentation. Sprague (1995) and Verderber (1991) are general-purpose texts on public speaking. Freeman et al. (1983) and Becker and Keller-McNulty (1996) are designed for speakers at professional meetings for statisticians. Regardless of the intended audience, each of these works and others that you will find in libraries and bookstores contain valuable advice for oral presentations and are worth reading.

Read about strategies for graphical presentations: Recognize that all graphical presentations are not equally effective. Cleveland (1994) and Tufte (1992) provide excellent strategies for making graphical presentations. As the field of statistical graphics is making rapid advances, we can expect to see major changes in the ways statisticians present data.

■ ■ ■ ■ ■ REFERENCES

Becker, Richard A., & Keller-McNulty, Sallie (1996), "Presentation Myths." *The American Statistician,* 50, 112–115.

Cleveland, William S. (1994), *The Elements of Graphing Data* (revised ed.). Summit, NJ: Hobart Press.

Freeman, Daniel H., Jr., González, María Elena, Hoaglin, David C., & Kilss, Beth A. (1983), "Presenting Statistical Papers." *The American Statistician,* 37, 106–110.

Sprague, Jo (1995), *The Speakers Handbook* (4th ed.). San Diego: Harcourt Brace Jovanovich.

Tufte, Edward R. (1992), *The Visual Display of Quantitative Information* (reprinted ed.). Cheshire, CT: Graphics Press.

Verderber, Rudolph F. (1991), *Essentials of Informative Speaking Theory & Contexts.* Belmont, CA: Wadsworth.

■ ■ ■ ■ ■ ORAL PRESENTATION CHECKLIST

This checklist gives reminders for preparing oral presentations.

Presentation Goal
___ Do you have a clearly defined goal or goals?

Audience
___ What are the backgrounds of the audience?
___ What are they likely to know about the topic?
___ What do they know about your background and training?
___ How big will the audience be?

Ground Rules
___ How much time is allocated for your presentation?
___ What is the size of the room?
___ Does the room have any special features?
___ What audiovisual equipment will be available?
___ Will there be a microphone?
___ When will questions be asked?
___ Will other speakers present related material before your presentation? If so, what will they cover?
___ Is a written version of the presentation needed? If so, when is it due?
___ Is a presentation title needed?

Presentation Outline
___ Have you developed an outline?
___ Are the major sections listed as major outline headings?
___ Does each major section have complete sentence subheadings?
___ Has an amount of time been allocated to each major section?
___ Is the total of the time allocations consistent with the ground rules?

Greeting and Introduction
___ Do you establish a spirit of goodwill?
___ Do you or the person introducing you establish your credibility?
___ Do you acknowledge the work of colleagues on the project?
___ Do you give the goal of the presentation?
___ Do you start with enthusiasm?

Preview
___ Do you list the main topics of the presentation?
___ Do you introduce verbal signposts?
___ Do you make a transition to the next section using its verbal sign-post?

Background
___ Do you give the background needed to understand the data collection activity?
___ Are you boring the audience with information they already know?
___ Do you explain why the activity is important?
___ Are visual aids used to explain products or complicated processes?
___ Is your background presentation nonstatistical?
___ Do you make a transition to the next section using its verbal sign-post?

Description of Data Collection Activity
___ If the project involves an experiment, do you explain the experiment?
___ If your project involves a survey, do you explain how the sample was selected and how the data were collected?
___ Are visual aids used to highlight key points of the design?
___ Does your description avoid jargon that the audience will not understand?
___ Do you make a transition to the next section using its verbal sign-post?

Results of the Data Analysis
___ Do you summarize the analysis in language the audience will understand?
___ Do you use visual aids to help explain the results?
___ Do you state the results in terms of the setting of the problem rather than in statistical jargon?
___ Do you make a transition to the next section using its verbal sign-post?

Conclusions and Recommendations

___ Do you summarize the goal, the data collection activity, and the major results?

___ Do you recommend specific actions to be taken?

___ Do you invite the audience to ask questions?

___ Do you finish with enthusiasm?

Presentation Mode

___ What presentation mode will you use? (Speaking from notes is generally best.)

Physical Delivery

___ Have you selected appropriate clothes?

___ Do you stand with your weight balanced on both feet?

___ Do you let your arms hang loosely at your sides except when gesturing?

___ Do you keep an expressive face complete with a smile during your presentation?

___ Do you maintain eye contact with the audience?

Oral Delivery

___ Are you enthusiastic?

___ Do you speak with enough volume to be heard at the back of the room?

___ Do you clearly enunciate each word?

___ Do you speak at a rate that can be comfortably understood?

___ Have you removed the clutter?

Presentation Notes

___ Have you prepared note cards?

___ Do you use key words rather than complete sentences?

___ Have you numbered your note cards?

Visual Aids

___ Can the visual aids be seen by the entire audience?

___ Are they legible?

___ Can they be easily read from the back of the room?

___ Can they be easily understood?

___ Do you use phrases rather than complete sentences in the visual aids?

___ Is there sufficient white space?

___ Do you effectively use color?

___ Have you practiced with the projection equipment?

___ Are the images aligned properly?

___ Are your visual aids in the proper order?

___ Have you proofread your visual aids?

___ Have you indicated on your speech notes where visual aids will be used?

___ Are you looking at the audience rather than your visual aids?

Handouts
___ What type of handout will you use?
___ Does your handout include your name, address, and other vital information?
___ How and when will the handouts be distributed?

Rehearsing
___ Are you starting your rehearsals several days before the presentation?
___ Are you beginning by reading the outline aloud several times?
___ Have you then rehearsed your entire presentation in private? If so, did you give extra attention to rough spots?
___ Have you given your entire presentation to a trusted friend?
___ Did you determine the amount of time used on each major section?
___ Have you fine-tuned your presentation to meet the time requirements?
___ Did you remove clutter?

Overcoming Fear
___ Are you comfortable about making the presentation? If not, have you analyzed your fears as specifically as possible? Have you tried replacing negative thoughts with positive ones? Have you tried relaxation techniques?

Question-and-Answer Period
___ Are you approaching this period with a positive attitude?
___ Are you willing to say "I don't know" if you don't know?
___ Are you understanding the question before you answer?
___ Are you collecting your thoughts before you answer?
___ Are you giving your answers to the entire audience?
___ Are you maintaining eye contact with the audience?

C H A P T E R
14

Producing Visual Aids
with PowerPoint

■ ■ ■ ■ ■ **INTRODUCTION**

PowerPoint software makes it easy to produce quality visual aids. This brief introduction will acquaint you with some PowerPoint features and allow you to produce some excellent visual aids. The version of PowerPoint that you use may work a little differently from that described here. If so, most modifications should be reasonably obvious.

■ ■ ■ ■ ■ **TEMPLATES**

PowerPoint offers a wide variety of visual aid templates. Each template gives a basic format for the visual aid. There are different template categories based on the intended presentation media: black and white overhead transparency, color overhead transparency, on screen, or 35-mm slides. The visual aid's style changes as you change the template. You will like some templates more than others. I like Double Lines. It is easy to change templates if you change your mind.

■ ■ ■ ■ ■ **SLIDES**

A PowerPoint document consists of one or more slides. These slides can be displayed on the monitor and printed. With the proper equipment, these slides can be turned into overhead transparencies and 35-mm slides or projected directly from the computer onto a movie screen. Each slide consists of a title and a body. The title contains the slide's subject or major point and the body contains supporting text or graphics.

▪ ▪ ▪ ▪ ▪ CREATING A POWERPOINT DOCUMENT

Double-click the PowerPoint icon. Then double-click the Blank Presentation button in the dialogue box that appears. You will be asked to select a layout for the first slide. Double-click the leftmost layout in the top row and blank slide 1 appears. Enter the first slide's title. This is often the presentation title. It appears in the title section of the slide as you type. Click in the body section of the slide. Enter your first supporting item. This is often your name. Press the enter (return) key before entering each additional supporting item.

To change the layout, click Layout at the bottom right of the window. A set of layouts appears in a dialogue box. Double-click the second layout in the top row. Notice that the layout of slide 1 has changed. The title has moved up and bullets now highlight the supporting items.

To create a new slide, click New Slide at the bottom right of the screen. After selecting a slide layout, enter the appropriate title and supporting information for the slide. You can move between the slides by clicking the arrows in the lower right of the window or by dragging the scroll bar.

You can change the font and character size, style, or color by selecting Font under the Format menu. Selecting Bullet under the Format menu allows you to change or eliminate the bullet symbol. Selecting Line Spacing under the Format menu allows you to adjust the amount of space between lines. To edit while in the slide view, click the desired location and then delete, insert, or highlight in the usual fashion.

▪ ▪ ▪ ▪ ▪ THE BUILD FEATURE

If you plan to project the computer screen image onto a movie screen, the build feature gives you the option of presenting each slide one line at a time. This has the advantage of presenting each line at the precise time you want the audience to see it. To use the build feature, select Build under the Tools menu. You will want to explore the various options. It is possible to highlight the new information and dim the previous lines on a slide.

▪ ▪ ▪ ▪ ▪ OUTLINE VIEW

In the slide view, you see one slide at a time. To see the text portion of all slides, select Outline under the View menu. Use the keyboard arrows or the scroll bar to move through a document. You can also create or edit slides in the outline view. The arrow icons in the upper left of the window allow you to move between a slide's title and body sections and to format items and subitems within the body section.

■ ■ ■ ■ ■ NOTES VIEW

In the notes view, you see a slide image in the upper half of the window and another section in the lower half. This lower section is designed for presentation notes. These notes appear only in the notes view and when the notes pages are printed. To enter the notes view, select Notes Pages under the View menu.

■ ■ ■ ■ ■ SLIDE SORTER

You may easily reorder your slides in the slide sorter view. Enter this view by selecting Slide Sorter under the View menu. Images of each slide will appear. Click and drag the slide's image to its new position. For example, if you wish slide 6 to come immediately after slide 2, click and drag the image of slide 6 between the images of slides 2 and 3. You can also duplicate or delete a slide by clicking on the slide's image and then selecting Duplicate or Cut under the Edit menu.

■ ■ ■ ■ ■ SLIDE SHOW

The slide show feature allows you to sequentially view all the slides on the monitor. With a special projection device, they can also be viewed on a movie screen. To use the slide show, select Slide Show under the View menu. You can have the computer change the slides after a predetermined number of seconds or manually change the slides by pressing any key on the keyboard or the mouse. You end the slide show presentation by pressing the escape key.

■ ■ ■ ■ ■ CHOOSING THE TEMPLATE

To choose or change the template, select Apply Design Template (or Presentation Template) under the Format menu. A list of templates will appear. Double-clicking a template name applies the template to all slides. Explore the various templates and pick your favorite.

■ ■ ■ ■ ■ PRINTING

You can print slides, note pages, outline view, or handouts. Handouts consist of several slide images per page. To print, select Print under the File menu and then select what you want printed from the Print What item in the Print window. Printing handouts with 6 slides per page allows you to assess if you have too much information on your slides.

■ ■ ■ ■ ■ ASSIGNMENT: PRACTICING WITH POWERPOINT

1. Create a PowerPoint document in which the first slide has the title "PowerPoint Practice," with your name and today's date in the body, and the second slide has the information from Figure 13.1.
2. Print the document as you would for making overhead transparencies.
3. Print a handout containing these two slides on a single page.

CHAPTER

15

Strategies for Effective Consulting

■ ■ ■ ■ ■ **INTRODUCTION**

Statisticians have three major roles in the workplace: researcher, teacher, and statistical consultant. While the time spent in each role varies greatly for different positions, most statisticians spend time in each role.

The statistician as researcher reads the literature (textbooks and journals) and attends professional meetings to learn about new developments in the field. He or she may also develop new statistical methods or investigate properties of existing methods. The proofs in mathematical statistics courses are examples of this type of research.

The statistician as teacher trains individuals in statistical thinking, statistical methods, and statistical theory. Statistics professors clearly perform this role, but they are not the only teachers. Most statisticians in government and industry spend part of their time training other employees.

You are probably familiar with the researcher role and the teacher role. However, you may not be as familiar with the statistical consultant role. The statistical consultant uses his or her statistical knowledge to help solve problems that arise in fields other than statistics. A key aspect of this role is that the statistician collaborates with nonstatisticians to solve their problems. These nonstatisticians are called clients.

Statistical consultants work on exciting problems from a variety of fields. One project may deal with medical research or engineering and the next with problems in the printing department. It can be fun to learn about and contribute to many different fields.

Consultants routinely use a variety of non-statistical skills, including skills in computing, oral and written communication, project organization, and working with a variety of personalities. We develop these skills through a combination of education and on-the-job experience. Boen and Zahn (1982) present an excellent discussion of the non-statistical aspects of statistical consulting.

This chapter introduces you to the role of the statistical consultant and gives strategies for success. It also discusses some of the ethical issues that statistical consultants face. Finally, it references additional materials on statistical consulting.

■ ■ ■ ■ ■ APPROACHES TO STATISTICAL CONSULTING

There are several ways to do statistical consulting—some are good and some are not. Let's consider three common approaches.

Hey buddy, do you have a minute? In this approach, the client sticks his or her head in the statistician's door or calls on the phone and says something like "Hey buddy, I know you're busy. Can you answer a quick question for me? I won't take much of your time." The request that follows is often similar to one of the following:

"Can you show me how to run a *t*-test on the computer?"
"My coworker did an analysis of his data. I used his program on my
    data. Can you tell me what this number on the printout means?"
"I computed this test statistic, but I don't have a table. Is it significant?"

This approach to statistical consulting is dangerous. In answering such a quick question, the statistician runs the risk of making what Kimball (1957) calls an error of the third kind, that is, giving the right answer to the wrong problem. If the statistician answers the question without getting more information, he or she may never understand what the real problem is or if the client's methodology is appropriate.

Many clients with only a little knowledge of statistics tend to view every problem as fitting one of the few terms they know. This may be happening in the first scenario with the request for the *t*-test. Also, several statistical procedures yield *t*-tests. Is this a one-sample problem, a two-sample problem, a regression problem, a test of independence, or something else? Are the assumptions satisfied? Do they have a one-sided or two-sided alternative? Clearly, the question is not as simple as it first sounds. Can you see similar concerns arising with the second and third scenarios?

This approach to consulting uses only a small part of the statistical consultant's expertise. The statistician did not take part in the experimental design, data collection, data editing, choice of analysis, or explanation of the results. Rather, he serves as what Bross (1974) calls a "shoe clerk." The client is not allowing the statistician to make his or her full contribution to the study.

Help! I'm going under for the third time. In this approach, the client comes to the statistician's office and says,

> "Please help me! My boss (advisor, editor, . . . , etc.) says I have to have some statistics in my report. I've never studied statistics, and numbers scare me. Please analyze these surveys, and assume whatever you need to. By the way, I need it tomorrow."

The difficulty in this approach is that the client is asking the consultant to do both the statistician's job and the client's job. The statistician, most likely, is not an expert in the client's field. The client needs to provide vital information before the statistician can choose the proper analysis.

As in the previous approach, the statistical consultant is being called in very late in the study. It is too late to make suggestions regarding the sampling plan or the wording of the survey instrument. Advice on these matters often greatly improves project quality.

A collaborating scientist: In this approach, the client contacts the statistical consultant at the beginning of the investigation. Working as a coworker, the statistician assists in problem definition, experimental design, sampling, data collection, data entry, data editing, data analysis, drawing conclusions, and preparing reports. He or she asks probing questions and gives the client advice at each stage of the study. Daniel (1969) and Deming (1965) give additional thoughts on the roles of the client and the statistical consultant in such collaborative efforts.

The collaborating scientist approach is the ideal. It allows the statistician to have the greatest impact on the investigation. However, you won't always have the luxury of entering a study at the beginning. In many cases, you will have to make the best contribution that you can to a study that has already begun. In such cases, politely show the client that you would have been more helpful if you had been in on the study from the beginning. Hopefully, the client will include a statistician in the planning stage of his or her next study.

▪ ▪ ▪ ▪ ▪ THE PROCESS OF STATISTICAL CONSULTING

Statistical consulting is done in different ways with different degrees of success. The following steps have been useful to me in collaborating with clients:

1. Give the client a friendly greeting and tell him or her that you are happy to be working with them.

Statistics and statisticians intimidate many clients. Also, you may be nervous working with a new person or on a problem you haven't seen before. A friendly greeting breaks the ice for both the client and the statistical consultant. Having the meeting in the client's office, rather than yours, may relax the client. It may also allow you to observe more about his or her project and to meet other people involved in the project.

2. Begin the first session by having the client give a summary of the study that he or she has planned or has already done.

Don't answer any specific questions until you understand the study setting! If you don't understand something, ask a question. If you still don't understand, ask another question. If you still don't understand, keep asking questions.

Clients often use jargon specific to their field. Don't hesitate to ask what terms mean. It is a big mistake to act as if you understand things when you really don't. It is difficult, if not impossible, to give good advice without understanding the study setting. Later in the consulting process, the shoe may be on the other foot. You may need to explain statistical terms to the client. Open communication is vital to successful statistical consulting.

Some clients try to speed up the process by describing their study in statistical terms. Encourage them to let you fully understand the study before you discuss an appropriate analysis.

Take careful notes throughout the consulting process. It is very easy to forget key details. This project may last for several months. You may start several other projects before this project ends.

3. Have the client state the goals of the study in nonstatistical terms.

The question "What are you trying to find out or prove in your study?" helps the client give you the necessary information. By stating the goals in nonstatistical terms, the client allows you to consider a wider range of possible analyses.

Some clients will need help in formulating their project goals. You may need to ask some questions to help them clarify their goals. It is a good idea to restate your understanding of the project goals to give the client the opportunity to clear up any misconceptions on your part.

4. Formalize the goals into statistical terms and decide on a tentative method of analysis.

In this step, you are translating from the client's jargon to the jargon of statistics. Now you may have to explain terms to the client.

Don't feel pressured to produce a complete analysis plan off the top of your head. Experienced statistical consultants often have to spend time thinking about a client's problem, talking with colleagues, and reading the literature before they choose a tentative analysis. If a tentative analysis is not obvious, tell the client that he or she has presented an interesting problem and you need to research the matter before you can give useful advice. You can then prepare the tentative analysis before the next meeting.

Even if you know the general approach, you may not know all the details. For example, you may know that you will regress one variable on another, but not know if you will need to transform the data. In another example, you may know that you will compare two treatments, but not know if an assumption of normality is reasonable.

In choosing the tentative analysis, don't underestimate the importance of relatively simple graphs and charts. Simpler analyses are more easily understood and have more impact than complicated ones. If a simple analysis will adequately tell the story, use it. You can also do a more complicated analysis to convince yourself that the simple one is telling the correct story.

Don't try to pound a square peg into a round hole. Make the tentative analysis fit the needs of the project rather than changing the project to fit a particular type of analysis. Remember, you are helping to solve the client's problem, not changing the problem to suit your needs.

Beginning statistical consultants often want to be sure that they come up with the very best possible tentative analysis. Often, you can't do this in a reasonable amount of time. A timely good analysis is often better than a very late optimal analysis.

5. Assist the client in designing the experiment.

In this step, you work with the client to get the most relevant information from the study. In a project involving surveys, you jointly review or develop the survey instrument and choose the sampling plan. In a planned experiment, you discuss the selection of factors, levels of factors, and the particular design to be used. You should address issues of sample size and its impact on the precision of estimators or the power of hypothesis tests. The client may need to rethink the scope of the study based on this discussion.

Research studies often require considerable time and money. You and the client need to be confident that the experiment will answer the important questions. Clients are very unhappy if they spend months of time and much money collecting data and then learn that they cannot estimate an important parameter, that a confidence interval is so wide that no meaningful conclusions can be made, or that the power of the hypothesis test was so low that it was almost impossible to have rejected the null hypothesis.

Sometimes, clients don't have the resources to collect the data needed to answer their important questions. Show them this before, rather than after, they spend their resources to collect the data.

6. Discuss how the data will be collected.

In this step, you want to explore the data collection plan for any possible biases that might invalidate the study.

7. Assist the client in developing data entry and data editing plans.

Although you are comfortable working with data, many clients are not. Developing a data entry plan, entering the data, and editing the data for errors often overwhelm clients. Work with the client to ensure that these tasks are competently done. At times, you will need to do this work or arrange for someone else to do it for the client. This step is crucial to the consulting process. You need good data to make good decisions. Most data sets contain some suspicious entries that need checking.

8. Assist the client in performing the analysis and making statistical conclusions.

Notice how late the data analysis comes in the statistical consulting process. If you participate in the study from the beginning, it may be several months after the first meeting before you analyze the data. The careful notes you have taken at each step will refresh your memory about the project.

It is now time to analyze the data. In some cases, the client will run the computer programs, and in other cases, you will do it. Either way, you want to carefully study the results. The client will need your help in determining things such as if assumptions are violated and if the model is appropriate. The client will also need help understanding the statistical conclusions.

9. Discuss the nonstatistical implications of the conclusions with the client.

Making the statistical conclusions doesn't finish the job. It is now time to translate these conclusions into the language of the client's field of study. After you have done this, have the client restate his or her understanding of the results. This gives you the opportunity to explore any misunderstandings that may have occurred.

Some clients tend to make stronger conclusions than are justified by their data. This discussion gives a good opportunity to explain exactly what can and can't be said based on the data.

10. Write a report summarizing what has been done and what the conclusions are.

Verbally interpreting the statistical results in the language of the client's field doesn't finish the job. You now need to describe the analysis in writing. The written report should include a description of the experiment, including the experimental design or sampling plan, the analyses that were performed, any assumptions and models used in the analyses, and the specific conclusions that were reached. You should also explain any limitations of the study or of the conclusions.

As a beginning consultant, you may think this step is not necessary. However, clients usually write reports about their studies. If they don't completely understand the statistical aspects of the study, their report becomes garbled. This makes you and the client look bad and weakens the study. Clients often don't admit that they don't understand the statistician's verbal conclusions. They also tend to forget details between meeting with you and writing their report.

11. If possible, read the client's report prior to its distribution.

Make sure the descriptions of the statistical aspects are accurate and the conclusions are consistent with the data. Imagine your dismay if the client's report thanks you for your assistance and then presents an incorrect or garbled description of what you did. After all the work you have done on the project, you end up the goat. This has happened to many statistical consultants.

You should offer, if not insist, on reading a draft of the client's report. If you have been a major contributor to the project, you should also be listed as a co-author. Negotiate this with the client at the first meeting.

■ ■ ■ ■ ■ ETHICAL CONSIDERATIONS IN STATISTICAL CONSULTING

You probably think of attorneys and medical doctors as professionals. They have professional organizations that certify their members as meeting certain minimum standards and codes of professional conduct. Members who violate these codes can lose their license to practice the profession.

Statisticians are also professionals. The American Statistical Association has a code of conduct but does not license or certify statisticians. *The Ethical Guidelines for Statistical Practice,* first published in 1989, gives a set of expected behaviors for statisticians. The American Statistical Association's web site (*http://www.amstat.org*) gives more information about these guidelines. As with attorneys and doctors, unethical conduct by a statistician damages the entire profession. This section will explore two of the common ethical dilemmas faced by statistical consultants—neutrality and confidentiality.

Let's first explore neutrality. In many statistics books, the decision between rejecting a null hypothesis and failing to reject a null hypothesis appears to be only a mathematical exercise. The same is true for determining the endpoints of a confidence interval. The reader does the calculations, and life goes on.

When the statistical consultant applies statistics to other fields, life is often not as simple. Large amounts of money or fame may rest on the outcome of statistical analyses. Consider the following three situations:

A pharmaceutical company may make millions of dollars from sales of a new drug if their statistical studies establish that the drug is safe and effective in treating a common disease. If their statistical studies don't establish that the drug is safe and effective, they make nothing. The researcher who developed the drug either gets high praise and a large bonus or goes back to the lab and tries again. Having to go back to the lab too many times between successes may jeopardize the researcher's employment.

A college professor may publish a scientific paper in a prestigious journal and make progress toward promotion and tenure if his or her statistical studies show significant results. If the statistical study does not show significant results, the paper often is not published.

A city will qualify for a major government grant if a survey of its residents shows certain results and gets nothing if the survey doesn't show these results.

In these situations, the client or sponsoring organization has a great deal to gain or lose depending on the outcome of the statistical analysis. Clearly, they strongly hope that the statistical results support their position.

The statistical consultant must be neutral. It is our job to make objective decisions in the face of uncertainty and emotion. Statisticians are the umpires of science. Just as a baseball game would quickly deteriorate with a biased umpire, the quality of science and the value of the statistical profession would quickly deteriorate if statisticians were not objective. Deming (1965) states that the statistician's attitude regarding an investigation should be "I couldn't care less" as to which side the results favor.

The statistical consultant can feel caught in the middle between the desires of the client and the ethical need to be neutral. This can be particularly problematic when the client is the consultant's boss or a powerful person in the organization. Gibbons (1973, p. 74) states, "It is essential that the statistician inform his employer of his neutral position on all strictly nonstatistical aspects of the study before agreeing to undertake an investigation, as his position as an independent agent is considerably weaker once the study commences." If your employer can't accept your neutrality, they are not allowing you to be a professional statistician. In that case, it is wise to consider finding a different employer. The statistics profession relies on the confidence of the public and the scientific community that we objectively analyze data.

It is the obligation of the statistician to design the study to get as much information as possible. We use the best designs, sampling plans, and estimators that we can. In this way, we maximize the probability that the client will be able to establish a particular "fact" if that "fact" is true. However, we must insist that good statistical practice is being used in the design and analysis of experiments, and we must be objective in drawing conclusions from the data.

Different professions have different codes of conduct. These codes may appear to conflict. For example, an attorney asked me to analyze some data for a court case. I told him that I would be happy to do so, but that I would have to be totally objective in my analysis. He indicated that he wanted an objective analysis. However, he would have me testify about the analysis only if it helped his client. This was perfectly reasonable because his code of conduct required him to make the best possible defense for his client. It would have been wrong for me to provide a nonobjective analysis, and it would have been wrong for the attorney to have presented less than the best defense for his client. In another case, he advised his client to settle when my analysis showed the data supported the other side. In both cases, the objective data analyses were valuable to the attorney and his clients. In legal cases, you may also be asked to evaluate the analyses of experts for the opposing side.

The second ethical issue we want to explore is confidentiality. Statisticians collect a large amount of sensitive data. These data can include information about people, such as their ages, weights, incomes, health histories, and opinions on controversial issues. It can also include information about businesses, such as the average number of hours worked by their employees, their plans to hire or layoff workers, and their plans for capital expenditures. The collection of this information is vital to social science, medical, and economic researchers in academics, business, and government.

These data are generally collected in surveys. The success of these surveys depends on the willingness of individuals and businesses to participate in the study and to provide accurate information. Implicit in this process is the respondents' assumption that their data will be confidential. That is, their data will be included in summary statistics and other statistical analyses, but their individual data will not be released. If the participant does not have this trust, then they will either not participate or give inaccurate data.

As a statistician, you should not release confidential information and you should do all in your power to keep others from doing so. One way of increasing confidentiality is to identify data records by a respondent number rather than by the respondent's name. This makes it much more difficult for nosy data snoops to link confidential information to specific respondents. The release of confidential information not only damages the respondent and your reputation, but it also makes it more difficult for researchers and statisticians to collect data in future studies. It may also be illegal.

▪ ▪ ▪ ▪ ▪ IMPROVING AS A STATISTICAL CONSULTANT

Most statisticians are not excellent consultants in their first attempts. The craft of statistical consulting takes time to master. You will want to improve your consulting skills throughout your career. As with most skills, there are three ways to improve: practice, observe, and read.

Practice: If possible, your first consulting efforts should be a joint effort with a more experienced consultant. Initially, you will probably observe more than you participate.

When you are ready to solo, it is best if your first clients have relatively simple statistical problems. This will allow you to build your confidence and pay more attention to the nonstatistical aspects of the consulting relationship. As you get more practice, you will become more comfortable in the consultant's role and be ready to handle more complicated statistical problems. Even the most experienced statistical consultants get problems they don't know how to handle. Don't be afraid to speak to colleagues about a tough problem.

Watch videos of some of your consulting sessions to see where you have done well and where you need improvement. Ask trusted mentors to evaluate your performance and tell you how they might have done things differently.

Observe other consultants: Different statistical consultants use different styles. You can learn about these styles by sitting in on their consulting sessions, by talking to them about consulting, and by listening to their presentations at professional meetings. Statisticians are not the only people who interact with others, so you can also learn by observing experienced consultants in other fields. In these interactions, observe aspects of the consultant's style that are particularly good and any aspects that detract from their ability to help their clients. Consider incorporating the good into your style, and avoid the bad. Pay special attention to the types of questions the consultants ask.

Read about statistical consulting: There are numerous works on statistical consulting in the literature. Boen and Zahn (1982) give excellent advice on the nonstatistical aspects of consulting. They include a very useful chapter on the management of a consulting practice. Deming (1965) and Gibbons (1973) provide excellent advice on ethics. Bross (1974), Deming (1965), and Daniel (1969) address the responsibilities of the client and the consultant. Woodward and Schucany (1977) give a bibliography of the papers on statistical consulting available at that time. They also list some of the numerous books on interpersonal relationships. Each of these authors presents valuable advice for statistical consultants.

■ ■ ■ ■ ■ REFERENCES

Boen, James R., & Zahn, Douglas A. (1982), *The Human Side of Statistical Consulting*. Belmont, CA: Lifelong Learning Publications.

Bross, Irwin D. J. (1974), "The Role of the Statistician: Scientist or Shoe Clerk." *The American Statistician, 28* , 126–127.

Daniel, Cuthbert (1969), "Some General Remarks on Consulting in Statistics." *Technometrics,* 11, 241–245.

Deming, W. Edwards (1965), "Principles of Professional Statistical Practice." *Annals of Mathematical Statistics*, 36, 1883–1900.

Gibbons, Jean D. (1973), "A Question of Ethics." *The American Statistician*, 27, 72–76.

Kimball, A. W. (1957), "Errors of the Third Kind in Statistical Consulting." *Journal of the American Statistical Association*, 52 , 133–142.

Woodward, Wayne A., & Schucany, William R. (1977), "Bibliography for Statistical Consulting." *Biometrics*, 33, 564–565.

CHAPTER
16

Strategies for Finding a Job

■ ■ ■ ■ ■ **INTRODUCTION**

Finding a job is a major marketing effort. You are marketing your skills to potential employers. You must learn which employers are hiring and convince them that you are the person they need. Your ability to sell yourself is key. You will usually be in competition with other applicants.

You can have a major advantage in this competition by being organized and paying attention to details. Take the initiative throughout your job search. You must make contacts to get interviews, and you must have interviews to get job offers.

Finding a job will take some time. You will apply for many jobs and probably be turned down several times. Don't be discouraged. There are many good jobs for statisticians, and you only need one.

This chapter presents information about the types of jobs that statisticians hold, the skills employers want, and strategies for getting hired.

■ ■ ■ ■ ■ **WHO HIRES STATISTICIANS?**

The federal government is the largest employer of statisticians. Examples of statistical projects include the Federal Census conducted by the Census Bureau, unemployment and inflation estimates conducted by the Bureau of Labor Statistics, crop yield estimates by the Department of Agriculture, monitoring of social programs by the Department of Health and Human Services, intelligence work by the Central Intelligence Agency, health research in the National Institutes of Health, the Centers for Disease Control and Prevention and the Food and Drug Administration, environmental research in the Environmental Protection Agency, and selection of individuals for audit by the Internal Revenue Service. There are also numerous state government positions.

Businesses also employ statisticians. Examples of statistical projects include experimental design and process monitoring to improve quality in manufacturing, experiments in the pharmaceutical industry to establish that new drugs are safe and effective, credit scoring algorithms to determine consumer credit worthiness, forecasting demand for goods and services, market research strategies including consumer satisfaction surveys, determining premiums for insurance companies, developing statistical computing packages, and writing sophisticated SAS programs.

Colleges and universities employ statisticians. Besides statistics professors, statisticians work in statistical consulting centers, in institutional research offices, and in educational testing organizations.

Jobs in government, business, and academics often require a combination of education and work experience. Many jobs require an advanced degree. If you have a solid academic record, consider going to graduate school. The American Statistical Association's web site (http://www.amstat.org) has a list of schools offering degrees in statistics. Most programs have many assistantships to support graduate students.

■ ■ ■ ■ ■ JOB TITLES

Statisticians have a large number of job titles. Many of these titles don't include the word *statistician*. Figure 16.1 gives a partial list of these titles. Searching a broad range of job titles will find many more openings.

Actuarial Trainee	Operations Researcher
Biostatistician	Planner
Data Analyst	Quality Engineer
Decision Scientist	Quantitative Analyst
Educational Analyst	Research Administrator
Forecasting Analyst	Research Analyst
Health Statistician	SAS Programmer
Industrial Engineer	Statistical Analyst
Marketing Analyst	Statistician
Mathematical Statistician	Survey Researcher

FIGURE 16.1 Some job titles for statistics jobs

■ ■ ■ ■ ■ SKILLS EMPLOYERS WANT

It should not surprise you that employers want statisticians to have good statistics and computing skills. However, they want more. Employers want enthusiastic, confident employees who complete tasks on time and clearly communicate the results. You need to be able to function on your own and as part of project teams. There is an emphasis on teamwork, oral and written communication, interpersonal skills, and organizational skills. Employers also expect all employees to be completely trustworthy. Evidence of previous work success is also desirable.

Your marketing goal is to convince employers that you have these skills. They will evaluate you every time they read something you have written, talk with you, or talk with one of your references. Be sure these interactions show you at your best. If you are weak in any area, strive for improvement.

■ ■ ■ ■ ■ WHEN TO START YOUR SEARCH

Start your job search about nine months before graduation. This allows you to participate in on-campus interviews for an entire academic year. If you missed this natural starting point, start as soon as possible.

■ ■ ■ ■ ■ WHERE TO START YOUR SEARCH

You will interact with many employers over several months. It is easy to forget important details. Start your job search by getting a three-ring binder and some paper. This will be your job search notebook. Keep careful notes, using a different page for each potential employer.

Your next step is to visit your campus career center or placement office. These offices assist students in finding employment. Their services generally include scheduling on-campus employment interviews, providing information on potential employers, and improving students' résumés and interviewing skills. On your first visit to the career center, find out the following:

How on-campus interviews are announced and how to sign up
What services they provide to critique your résumé and interviewing skills
How to locate information about potential employers
How to access databases listing job openings in your field

Unless your school graduates a large number of statistics majors, the number of on-campus interviewers may be small. That's okay. The career center is only one of the resources you should use in your job search.

■ ■ ■ ■ ■ **JOB TARGETS**

The next step in your job search is establishing job targets. What geo-graphical constraints do you have and what type of job do you want?

Restricting yourself to a particular location greatly reduces the num-ber of potential employers. Don't be too restrictive unless you absolutely have to be. There are many great places to live.

Many students start the job search without having thought about the type of work they want to do. This leads to a résumé that is not targeted toward any specific group of employers. It is better to produce a separate résumé for each targeted group. For example, you might have one résumé for use with pharmaceutical companies, another for positions in quality improvement, and a third for positions as a survey statistician. These résumés will be similar, but they are each designed to appeal to a targeted group of employers.

■ ■ ■ ■ ■ **REFERENCES**

At some point in the hiring process, employers will contact your refer-ences. The references will evaluate you in terms of the skills the em-ployer wants. Select three or four credible people who are familiar with your skills and who will say good things about you. Professors familiar with your technical abilities are common references. If you have had technical work experience, your boss would be a good reference. Don't use family members or neighbors as references.

Always ask people if they are comfortable being a reference before listing them. If they express reservations, use another person. Provide your references with copies of your résumés and give them periodic up-dates on your job search. Always contact your references as soon as pos-sible after good interviews. The employer may be contacting them in the near future.

■ ■ ■ ■ ■ **RÉSUMÉ**

Your résumé is an advertisement of your abilities. The résumé's purpose is to persuade employers to interview you. Employers read résumés very quickly. The résumé you spent hours to create will be read in 30 seconds, perhaps by a computer! Design your résumé so that employers will quickly find evidence of the skills they want. Leave plenty of white space to facilitate easy reading.

You have one page to present your ad. Emphasize your strengths and accomplishments. Don't waste space on weaknesses. Make each word count—use phrases rather than complete sentences. Use action verbs (such as created, designed, planned, developed, improved, and implemented) to describe your accomplishments in coursework, jobs, and organizations.

Your résumé should include your name, address, phone number, and e-mail address. If you will move on a specific date, include information for before and after that date.

Include a professional objective. Examples are entry-level position as a statistician in the pharmaceutical industry and mathematical statistician involved in survey research.

Describe your educational background: colleges attended, date of attendance, degrees received, and majors and minors. List the degree you are about to receive as anticipated. If you have good grades, list your grade point ratio. If your grades in statistics or in the mathematical sciences are higher than your overall grades, also give that grade point ratio. A brief description of your program of studies, such as "program emphasized a balance of application and theory with a strong computing component," is helpful. If you wrote a thesis, indicate your thesis title and advisor. List any nonthesis publications in another section.

Describe any technical work experience. List the employer, job title, dates of employment, and a description of your accomplishments. Remember to use action verbs.

Give your country of citizenship. If you are not a U.S. citizen, indicate that you are a permanent resident or give your visa status.

Include the names, addresses, phone numbers, and e-mail addresses of your references. Reference information is usually given in the résumé's last section or on a separate page.

Select material for the rest of your résumé that emphasizes your strengths toward the specific job target. Possible items include honors, leadership and teamwork experience, special computing or foreign language skills, relevant coursework, membership in professional organizations, nontechnical work experience, and activities. If you worked to pay for a significant part of your college expenses, indicate the percentage of college expenses earned. Height, weight, gender, and health status should not be included. You can include a list of one or two hobbies to show that you are well-rounded.

Producing a good résumé requires careful editing and proofreading. Expect to make several drafts. When you have a good résumé, have the career center staff and one of your professors review it. You might also get comments from a friend in the business world.

■ ■ ■ ■ ■ SALARY EXPECTATIONS

Have a good idea of the salary you can expect in particular markets and industries before meeting with employers. While statisticians' salaries tend to be high relative to many professions, there is considerable variability. Part of this variability is due to cost-of-living differences. There are also differences in fringe benefits, such as health insurance, retirement, etc.

Get an idea of current salaries by talking with the career center staff, faculty, and recent graduates. You can also look at salaries quoted in job ads. Have a salary range rather than a specific number. The *Amstat News*, a publication of the American Statistical Association, publishes a yearly salary survey for statistics faculty positions.

■ ■ ■ ■ ■ LEARNING WHO IS HIRING

After thinking about your job targets, obtaining references, preparing your résumé, and learning about salary ranges, you are ready to talk to employers. You must now determine who is hiring. There are several ways to learn of openings.

Visit the career center. They schedule on-campus interviews and often have access to many other sources of job openings. Career centers offer valuable assistance. Contact them regularly to learn about upcoming interviews and seminars. On-campus interviewers can talk with only so many people. Take the initiative. Get your name on the list for employers hiring in your field.

Network with faculty, alumni, friends, and statisticians. Make sure everyone you know knows you are seeking employment. These people often know about openings or know people who do. Chapter meetings of the American Statistical Association and the American Society for Quality offer excellent networking opportunities. Don't be a wallflower. Talk to everyone at the meeting.

Surf the Internet. Many employers list openings on the Internet. The following sites are good sources: Monster Board (*http://www.monster.com*), Online Career Center (*http://www.occ.com*), HeadHunter.Net (*http://www.headhunter.net*), the University of Florida Department of Statistics (*http://stat.ufl.edu*), and Federal Career Opportunities (*http://www.fedjobs.com*). Federal Career Opportunities charges a fee. All others are free. Your career center staff may know of other sites. Many employers also announce openings on their web sites. Remember to search using several job titles.

Read periodicals. The *Amstat News* publishes job announcements in each issue. The American Statistical Association's web site displays some of these ads. You can join the Association as a student member and receive the *Amstat News* for a modest amount. Large city newspapers and the *National Business Employment Weekly* also list some openings for statisticians. Your library or career center may have these periodicals.

Contact employment agencies. Several employment agencies specialize in matching statisticians with potential employers. Some of these agencies advertise in the *Amstat News*. These agencies are best for placing students with advanced degrees and bachelor's-level students seeking jobs as SAS programmers. You should not have to pay a fee to use such an agency.

Go to the Joint Statistical Meetings. The Joint Statistical Meetings are the annual meetings of several statistical societies. The meetings occur in a different city each August. The meetings offer a placement service to match applicants and employers in a setting similar to on-campus interviews. You must register for the meetings to use the placement service. The American Statistical Association's web site gives information about the Joint Statistical Meetings.

Conduct a direct marketing campaign. In this approach you contact employers not advertising positions. There are thousands of employers, so you need to select those most likely to have positions matching your job targets. There are several ways to do this. The *College Placement Annual*, which should be in the career center, lists employers who hire in specific majors. The American Statistical Association's web site lists its corporate members. Your career center or library has lists of companies in particular businesses. If you have targeted a particular city, talk to the Chamber of Commerce and to members of the American Statistical Association living in that area. The American Statistical Association's *Directory of Members* lists members by city and state. Faculty members often have this directory.

Take the initiative—use as many ways to find openings as you can. On finding an opening, copy the important information in your job search notebook.

Mail a cover letter and your résumé to the potential employer unless an interview is arranged through the career center, employment agency, or placement service. Address the cover letter to a specific individual. The employer's telephone receptionist can give you the appropriate person's name and address. The one-page cover letter should introduce yourself, state that you are applying for a position, summarize your important qualifications, state how your skills can help the employer, and request an interview. Include the job title, job announcement number, and how you learned of the opening, if applicable. If you learned of the position through someone who the addressee knows, mention that person's name in the first sentence.

The cover letter is at least as important as the résumé. The cover letter provides the employer's first impression of you and helps route your résumé to the correct people. It may also convince an employer to create a position that will utilize your skills. Several drafts may be required before mailing your first letter. Edit the letter to remove unnecessary clutter, and proofread it carefully. Have your final draft reviewed by the career center staff or a faculty member.

■ ■ ■ ■ ■ BE EASY TO CONTACT

Many college students are difficult to contact. Make it easy for employers to contact you. Use an answering machine to take telephone messages. Your greeting on the answering machine represents you to all who call. Project a professional image. Employers are turned off by greetings suggesting that you are a party animal. Check your answering machine, mail, and e-mail at least daily. Give a well-prepared, prompt response to all messages.

■ ■ ■ ■ ■ INTERVIEWS

There are two types of interviews: screening interviews and job-site interviews. Screening interviews take place in career centers, in placement services, and on the telephone. Employers use short screening interviews to select finalists. The finalists are invited to the job site for a longer interview. The employer usually pays your expenses for an out-of-town job-site interview. Your purpose in the screening interview is to get a job-site interview. Your purpose in the job-site interview is to get a job offer.

Prepare carefully for each interview. Learn about the employer and the job before the interview. The employer's web site, your career center, library, and faculty may have information. Review the job ad and the list of skills employers want.

The first question is almost always a broad question asking you to talk about yourself. Use this question as an opportunity to present a 90-second commercial describing where you are from, where you went to school, your academic and work accomplishments, and your teamwork and leadership strengths. Anticipate other questions that you might be asked. Figure 16.2 lists some typical questions. Prepare answers that emphasize your strengths and show the skills employers want.

Can you tell me a little about yourself?
Why are you interested in this position?
What are your strengths and weaknesses?
What have you learned from your mistakes?
Do you have experience working in teams?
What are your career goals?
What would you consider to be an ideal job?
What special skills do you have?
How could you make this employer more successful?
Why should we hire you?

FIGURE 16.2 Questions you might be asked by an interviewer

How does this position support the mission of the employer?
How did this job opening occur?
What might be a typical first project in this position?
What is the anticipated career path for the person hired into this position?
Does this position offer the opportunity for me to grow professionally?
How frequently are job evaluations and salary reviews done?
Where is the home base for this position?
What is the next step in the hiring process after this interview?

FIGURE 16.3 Questions you might ask an interviewer

Have questions to ask the interviewer. Phrase the questions in ways that show you have learned about the employer before coming to the interview. Figure 16.3 lists questions you might ask. Good interviewers will answer some of these questions during their initial remarks.

Interviewing skills are improved with practice. Participate in practice interviews if you have the chance. An experienced interviewer can give you valuable advice. You will become more relaxed after a few interviews.

Dress neatly for the interview. Conservative suits with appropriate shoes are always good for interviews. Arrive 15 minutes early to gather your thoughts. Bring a pen, notepad, and at least three copies of the appropriate résumé.

The beginning of the interview is the most stressful part. Get the interview off to a good start by looking the interviewer in the eye, smiling, introducing yourself, and shaking his or her hand. Make sure you understand how to pronounce the interviewer's name. The interviewer's first impression of you is very important.

Maintain eye contact and smile throughout the interview. Be enthusiastic and confident, but not arrogant. Give answers that emphasize your accomplishments and your ability to work with others. Don't volunteer negative information. If negative information comes out, don't complain, give excuses, or lie. Explain what you have done to improve. Avoid one-word or rambling answers.

The interviewer is much more interested in the employer's needs than in your needs. Focus your answers around how you can help the employer.

Don't bring up salary. If your expectations are higher or lower than the employer's, you may hurt your chances. The best time to discuss salary is after the employer decides they want to hire you. They may then be willing to pay more than they had planned.

If an interviewer asks your salary expectations, give a salary range. State that other factors, such as the working environment, opportunity for professional growth, benefits, and cost of living, are also important to you.

The interviewer will make it clear when the interview is over. Thank the interviewer for the opportunity to discuss the position and ask for his or her business card. If you are interested in the position, tell the interviewer that you like what you've heard. If this is a screening interview, state that you are interested in visiting the job site to learn more about the position. If it is a job-site interview, state that you would be happy to provide any additional information that they might need to make their hiring decision.

There are two things to do after each interview. First, summarize the interview in your job search notebook. Include the date, the interviewer's name and job title, details about the job, and the main topics of the interview. This information will help you prepare if there are additional interviews.

Second, write a letter to the employer thanking them for the opportunity to discuss the position. Include the date, site of the interview, and the job title you interviewed for. Address this letter to the interviewer, if possible. Confirm your interest in the position and summarize how your training and experience would be useful in the position. Repeat your desire for a job-site interview or to provide any additional information. Mail this letter on the day of the interview.

▪ ▪ ▪ ▪ ▪ JOB OFFERS

Soon after an employer decides to hire you, they will make a job offer. This offer should provide the job title, salary and benefits for the position, the starting date, and a deadline for acceptance. If the position is in another city, the offer may provide a relocation allowance and moving expenses. If you receive the offer by phone, thank the caller, ask for written confirmation of the offer, and ask for time to consider the offer. There can be major problems if what you thought you heard does not match what the employer thought they said. If you don't understand the terms of the offer or if expected items are missing, call the employer to clarify any issues. If time is of the essence, you can probably arrange to have the written offer faxed to your school or sent by overnight express. There is no standard amount of time for acceptance deadlines. Two weeks from the offer's mailing date is frequently used.

The plot thickens when you receive an offer. You may accept, decline, negotiate for better terms, ask for a deadline extension, or attempt to get offers from other employers before your deadline. These decisions can be difficult. You may wish to talk with your family, the career center staff, and the faculty. Don't let a deadline pass without making a response.

It is difficult to evaluate an offer without knowing what other offers you might get. If the offer is one that you would seriously consider, immediately inform all interested employers that you have an offer and must respond by a specific date. Employers must then decide to move rapidly toward making an offer or forget you. You may be able to get a one- or two-week extension of your deadline if you anticipate receiving other competitive offers or if it is several months before the offer's starting date.

Negotiating for better terms will sometimes get you a better offer. Having another offer greatly enhances your negotiating position. However, employers may not have any flexibility in their offers. Aggressively demanding more when the employer has no flexibility can break down the negotiations. If an offer is amended, get it in writing.

Once you make a decision, call the person who made you the offer and send a letter announcing your decision. If you accept the offer, call again in a few days to confirm that they received your letter. Also, notify other interested employers that you have accepted a position.

■ ■ ■ ■ ■ IF YOU DON'T HAVE A JOB WHEN YOU GRADUATE

If you don't have a job when you graduate, your job is finding a job. Check every day for new openings on the Internet, in newspapers, and in other publications. Follow up on these positions and on positions that you previously applied for. Also, continue talking with the career center, employment agencies, faculty, alumni, and family friends. They often learn of unadvertised openings.

Don't take a long vacation before continuing your job search. This lets your leads get cold and makes finding a job harder. There should be time for a vacation between accepting a job offer and starting work.

■ ■ ■ ■ ■ JOB SEARCH CHECKLIST

This checklist gives reminders for conducting your job search.

Preliminaries
___ Do you understand the types of employers that hire statisticians?
___ Do you understand the various job titles that statisticians have?
___ Do you understand the skills that employers want?

Job Search Notebook
___ Do you have a job search notebook?
___ Are you using a separate page for each potential employer?

___ Are you keeping important information about each opening you find?

___ Are you keeping a summary of each interaction with potential employers?

Career Center

___ Do you know how on-campus interviews are announced?

___ Do you know how to sign up for on-campus interviews?

___ Will the career center critique your résumé?

___ Will the career center arrange a practice interview?

___ Do you know how to locate information about potential employers?

___ Do you know how to access databases that list jobs for statisticians?

Job Targets

___ Do you have geographic constraints? Are they more restrictive than they absolutely have to be?

___ Do you have specific types of jobs in mind? Are you willing to consider other types of jobs?

References

___ Have you selected three or four credible people to be your references?

___ Are they familiar with your technical abilities?

___ Are they familiar with your communication skills?

___ Are they familiar with your interpersonal skills?

___ Have you asked them if they are comfortable being listed as a reference?

___ Have you given them copies of your résumés?

___ Are you giving them periodic reports on your job search?

___ Are you contacting them after each good interview?

Résumés

___ Have you considered having more than one résumé?

___ Can an employer easily read your résumé in 30 seconds?

___ Have you used phrases rather than complete sentences?

___ Have you used action verbs?

___ Have you included your name, address, phone number, and e-mail address? If you will move upon graduation, have you included two sets of information?

___ Have you included a professional objective specific to your job target?

___ Have you described your educational background? Have you listed your upcoming degree as anticipated? If you have good grades, have you listed your grade point ratio?

___ Have you listed any technical work experience and your accomplishments?

___ Have you given your country of citizenship? If not a U.S. citizen, have you indicated permanent resident or visa status?

___ Have you included names and contact information for your references?

___ Have you selected additional material to emphasize your strengths?

___ Have you carefully edited and proofread your résumé?

___ Has the career center staff or a faculty member reviewed your résumé?

Salary Expectations

___ Do you have a good idea of the salary range you should expect?

___ Do you understand that some cities are more expensive to live in?

___ Do you understand that fringe benefit packages can be quite different?

Learning Who Is Hiring

___ Do you understand the various ways to learn of openings?

___ Are you using several of these ways?

___ Are you regularly visiting the career center to learn of upcoming interviews?

___ Are you searching any appropriate job databases in the career center?

___ Are you regularly searching sources other than the career center?

___ Are you responding to openings with a carefully written cover letter and résumé? Have you taken as much care with your cover letter as with your résumé?

Be Easy to Contact

___ Do you have an answering machine on your telephone? Does the answering machine greeting project a professional image?

___ Are you checking your answering machine, mail, and e-mail at least daily?

___ Are you giving well-prepared, prompt responses to all employer contacts?

Interviews

___ Are you learning about the employer and the job before the interview?

___ Does your appearance look professional?

___ Have you prepared a 90-second commercial about yourself?

___ Have you thought through answers to anticipated questions?

___ Do you have questions to ask the interviewer?

___ Are you smiling and maintaining eye contact with the interviewer?

___ Are you focusing on the needs of the employer?

___ Do you know the interviewer's name and have his or her business card?

___ Are you waiting for the employer to bring up salary?

___ Have you thanked the interviewer for the opportunity to discuss the job?

___ Have you expressed interest in having a job-site interview?

___ Have you summarized the interview in your job search notebook?

___ Have you written a letter thanking the employer for the interview and highlighting your strengths?

Job Offers

___ Is the offer in writing?

___ Are the job title, salary, benefits, starting date, and acceptance deadline stated?

___ Do you understand all parts of the offer?

___ Have you informed other interested employers that you have an offer with a deadline?

___ Have you responded to the offer by phone and by letter?

If You Don't Have a Job When You Graduate

___ Are you checking for new openings daily?

___ Are you following up on jobs that you have applied for?

___ Are you regularly contacting the career center, employment agencies, faculty, alumni, and family friends?

Index